Contents

	Introduction to the Events Management Theory and Methods Series	vi
	Preface	viii
	About the author	x
1	**Introduction, Concepts and Definitions**	**1**
	Data, information, and knowledge	2
	Explicit and tacit knowledge	5
	Individual, group and organisational knowledge	6
	Knowledge management	8
2	**Knowledge Management Challenges in Event Organisations**	**15**
	The 'pulsating' nature of events	17
	Knowledge management in project-based organisations	19
	Lack of time and resources	22
	Lack of expertise and trust among the team	23
	Large number of volunteers	25
	Creative, operational and strategic knowledge	27
	The issue of unlearning, forgetting and knowledge leakage	29
	The issue of knowledge hiding and knowledge hoarding	30
	Reinventing the wheel	33
3	**Knowledge Management Activities, Models and Frameworks**	**39**
	Key knowledge management activities	41
	Knowledge management frameworks	53
	Objectivist vs. practice-based perspective on knowledge management	66
4	**Relational and Practice-Based Knowledge Management**	**70**
	'Knowing', 'know-how', embedded and embodied knowledge	72
	Formal and informal knowledge practice rituals	74
	The role of emotions in knowledge practices	78
	Communities-of-practice	79
5	**Structural Elements of Knowledge Management**	**94**
	Human resource management and knowledge management	95
	Top-down, bottom-up and middle-up-down knowledge management	100
	Interdisciplinary teams and pods	103
	Knowledge management roles and responsibilities	106

6	**Cultural Elements of Knowledge Management**	**113**
	Organisational identity	114
	Organisational culture	116
	Motivation	123
	Trust	125
	Collaboration and co-creation	127
7	**Power and Knowledge**	**134**
	Power, politics and conflict in organisations	136
	Power as a resource	137
	The concept of power/knowledge	142
	Power and empowerment	144
8	**Appreciative Sharing of Knowledge**	**155**
	Appreciative Inquiry – making visible what works in an organisation	156
	Appreciative Sharing of Knowledge: what makes people share knowledge?	160
	Using stories and storytelling to share knowledge with others	163
9	**Practical Implications and Recommendations for Event Organisers**	**171**
	Knowledge management recommendations for event organisers	172
	Future research into knowledge management in event organisations	180
	Glossary	**185**
	Index	**189**

List of Figures

1.1: Data – information – knowledge.	2
1.2: Dimensions and levels of knowledge.	10
2.1: 'Pulsating' event organisations.	17
2.2: Knowledge management challenges in 'pulsating', project based event organisations.	19
3.1: Spiral evolution of knowledge conversion.	56
3.2: The four major knowledge flow functions applied to events.	59
4.1: Degrees of community participation.	82
4.2: Levels of participation at CMF.	*88*
5.1: Human resource management and knowledge management.	96
5.2: Interdisciplinary pod structure at QMF.	105
6.1: Categories of organisational culture.	117
7.1: Types of empowerment.	145
8.1: The four steps in Appreciative Inquiry.	158
8.2: Cycle of Appreciative Sharing of Knowledge.	161
9.1: Knowledge management recommendations for event managers.	173

List of Tables

1.1: Properties of explicit and tacit knowledge.	6
3.1: Knowledge management activities.	42
3.2: Four modes of knowledge conversion in events.	57
4.1: Differences between formal work groups and communities-of-practice.	80
5.1: Comparison of top-down, bottom-up and middle-up-down approaches to knowledge management.	102
7.1: Eight different power bases.	140
9.1: Past and future research into knowledge management in event organisations.	180

Introduction to the Events Management Theory and Methods Series

Event management as a field of study and professional practice has its textbooks with plenty of models and advice, a body of knowledge (EMBOK), competency standards (MBECS) and professional associations with their codes of conduct. But to what extent is it truly an applied management field? In other words, where is the management theory in event management, how is it being used, and what are the practical applications?

Event tourism is a related field, one that is defined by the roles events play in tourism and economic development. The primary consideration has always been economic, although increasingly events and managed event portfolios meet more diverse goals for cities and countries. While the economic aspects have been well developed, especially economic impact assessment and forecasting, the application of management theory to event tourism has not received adequate attention.

In this book series we launch a process of examining the extent to which mainstream theory is being employed to develop event-specific theory, and to influence the practice of event management and event tourism. This is a very big task, as there are numerous possible theories, models and concepts, and virtually unlimited advice available on the management of firms, small and family businesses, government agencies and not-for-profits. Inevitably, we will have to be selective.

The starting point is theory. Scientific theory must both explain a phenomenon, and be able to predict what will happen. Experiments are the dominant form of classical theory development. But for management, predictive capabilities are usually lacking; it might be wiser to speak of theory in development, or theory fragments. It is often the process of theory development that marks research in management, including the testing of hypotheses and the formulation of propositions. Models, frameworks, concepts and sets of propositions are all part of this development.

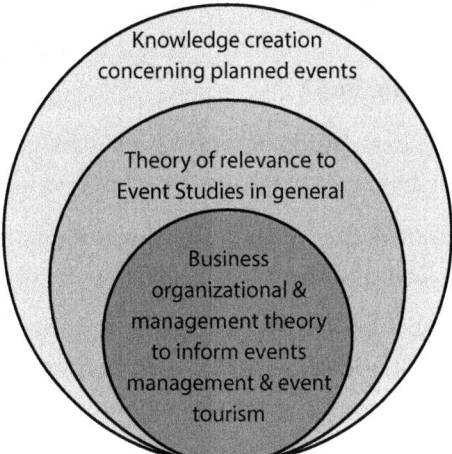

The diagram illustrates this approach. All knowledge creation has potential application to management, as does theory from any discipline or field. The critical factor for this series is how the theory and related methods can be applied. In the core of this diagram are management and business theories which are the most directly pertinent, and they are often derived from foundation disciplines.

All the books in this series will be relatively short, and similarly structured. They are designed to be used by teachers who need theoretical foundations and case studies for their classes, by students in need of reference works, by professionals wanting increased understanding alongside practical methods, and by agencies or associations that want their members and stakeholders to have access to a library of valuable resources. The nature of the series is that as it grows, components can be assembled by request. That is, users can order a book or collection of chapters to exactly suit their needs.

All the books will introduce the theory, show how it is being used in the events sector through a literature review, incorporate examples and case studies written by researchers and/or practitioners, and contain methods that can be used effectively in the real world.

Preface

Key objectives of the book

- Inform researchers, students and event managers on knowledge management and knowledge practices applied to events.

- Introduce knowledge management concepts, frameworks and models that can be adapted to a range of event organisations.

- Provide a textbook for event management students with study and discussion questions at the end of each chapter, as well as a useful resource and reference for event practitioners.

- Utilise case studies to explore theories and illustrate key points.

- Connect readers to the research literature through the provision of additional readings.

Organisation of the book

The first two chapters of this book provide an introduction to key terms and definitions, and then highlight specific knowledge management challenges for event organisations, such as the 'pulsating' nature of events and the lack of time and resources for knowledge management. Chapter 3 explores knowledge activities and knowledge management frameworks and models that can be applied to events, while Chapter 4 introduces the practice-based understanding of knowledge management and also explains how this can be used in an events context. Important structural and cultural elements of knowledge management are discussed in Chapter 5 and 6 respectively. The question of power and knowledge is explored further in Chapter 7, while Chapter 8 introduces Appreciative Sharing of Knowledge as an alternative approach to knowledge management. Lastly, Chapter 9 provides practical implications and recommendations for event organisers, as well as suggestions for future research on knowledge management in event organisations.

Stated at the beginning of each chapter are Learning Objectives, while Study and Discussion questions are presented at the end of each chapter. These are intended for students to apply some of the concepts and theories to their own event examples, and to reflect on their experiences of working for or volunteering at events. They can also be used as exam or essay questions. Suggestions for additional readings are provided at the end of each chapter, where some of the key elements of knowledge management have previously been researched in an events context.

The Queensland Music Festival case study has been incorporated throughout some of the chapters of the book to present best practice examples of how to effectively and efficiently manage knowledge in a festival organisation.

About the author

Raphaela Stadler is a Senior Lecturer in Event Management at the University of Hertfordshire, U.K. Her PhD (Griffith University, Australia) investigated different relational knowledge management practices within a festival organisation in Australia, and she has published numerous journal articles and book chapters on knowledge management/transfer in festival organisations, knowledge management rituals, power and knowledge, as well as community empowerment and cultural development. More recently, Raphaela has been interested in events, festivals and questions of well-being and quality-of-life. In this context, she is currently researching event attendance and family quality-of-life, and the impact of event and tourism experiences upon families with children with autism and children with type 1 diabetes. She is also involved in research on arts participation and memory creation in order to combat loneliness and isolation amongst people over 70.

1 Introduction, Concepts and Definitions

Learning objectives

- Learn key terms, dimensions and definitions of knowledge management.
- Understand different types and different levels of knowledge.
- Explore the relationship between knowledge management and organisational learning.
- Be able to define knowledge management and distinguish the three generations of knowledge management research and practice.

Introduction

During the late 1990s and early 2000s, the belief in a knowledge-based economy has grown; not just amongst academics, but also policy makers, consultants and managers. Nonaka and Takeuchi's (1995) work *The Knowledge Creating Company* was among the first to recognise that organisations that manage their knowledge efficiently, have a competitive advantage over organisations that do not succeed in doing so. Based on this understanding, a number of knowledge management frameworks and models have emerged which highlight how to improve the identification, creation, transfer, and documentation of knowledge. These will be discussed further in Chapter 3.

This introductory chapter starts with a definition of key concepts and terms, including data, information and knowledge; explicit and tacit knowledge; and the three levels of where knowledge resides (the individual, group, and organisational level). It also briefly explains how processes of managing knowledge at an organisational level can help organisations learn over time, create an organisational memory, and build on what has or has not worked in the past. The concept of knowledge management is thus linked to organisational learning and innovation (Argyris & Schoen, 1978; Gorelick et al., 2004; Senge, 2006). The final section of the chapter provides a range of knowledge management definitions and an overview of the 'three generations' of knowledge management.

Data, information, and knowledge

The knowledge management literature commonly distinguishes between data, information, and knowledge (see Figure 1.1):

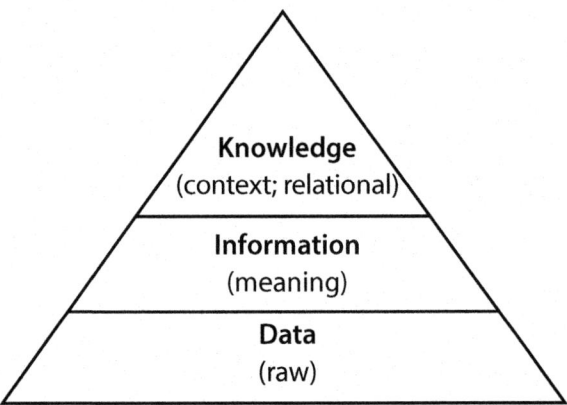

Figure 1.1: Data – information – knowledge

At the lowest level, *data* simply consists of (raw) numbers without any particular meaning; it can be described as discreet, objective facts. For example, the price for a ticket to the event, and when and where it was purchased. This kind of data does not tell us anything about why somebody bought the ticket, what their motivations were, if they were willing to pay a higher price, or who they are planning to attend the event with. Data can be stored on a computer though, and is of use to, for example, the finance, accounting and marketing team. Storing too much data, however, can be dangerous in the sense that the organisation might then struggle to make sense of the kind of data that actually matters and get lost in other non-relevant data. Furthermore, there is no meaning in the data, no interpretation, and it only partly describes what happened; it does not tell us how to make a decision based on this data.

As soon as data is put into a certain context or arranged in a specific order, it can then be turned into *information*. Information is "data that makes a difference" (Davenport & Prusak, 1998: 3); it shapes the person who receives it, it changes their way of thinking and their outlook in some form. It can be (see Davenport & Prusak, 1998: 3):

- **Contextualised**: what purpose was the data gathered for?
    - ◊ e.g. to find out people's motivations to attend an event
- **Categorised**: what are the units of analysis or key components of this data?
    - ◊ e.g. ask people about their needs, wants, and expectations of the event on a Likert scale
- **Calculated**: has the data been mathematically or statistically analysed?
    - ◊ e.g. analyse the data in terms of how many people strongly agree – agree – disagree – strongly disagree with certain statements
- **Corrected**: were there errors in the data and have they been removed/sorted out?
    - ◊ e.g. did somebody tick two boxes and has their answer therefore been removed from the data?
- **Condensed**: has the data been summarised to make sense of it?
    - ◊ e.g. produce figures and graphs to summarise findings about people's motivations to attend an event

Information can therefore also – to some extent – be captured by and stored on a computer, but it always requires human beings to first interpret the data. When information is interpreted further and put into action, it can then, at the highest level, be turned into *knowledge*; in other words, "knowledge derives from minds at work" (Davenport & Prusak, 1998: 5). Knowledge consists of experience, values, expert insights and contextualised information that is being put into practice by human beings. This happens through the use of information in, for example, processes, procedures, documents or repositories; and it adds value to the resulting activity of an individual, team or organisation (Du Plessis, 2006). Just as information derives from data, knowledge derives from information through processes such as: (Davenport & Prusak, 1998: 6)

- **Comparison**: how does information about the event compare to other information about the event that we already have?
  - e.g. how do people's motivations to attend the event compare to what they actually do at the event? Where do they go? What activities do they take part in? What do they buy? Are they repeat visitors or one-off attendees?
- **Consequences**: how can we use this information to be able to make decisions?
  - e.g. if low ticket prices turn out to be a key motivating factor for people to attend the event, shall we provide special discounts for, for example families, students, pensioners?
- **Connections**: how does this knowledge relate to other knowledge?
  - e.g. if we offer discounts, do we need to cut costs elsewhere?
- **Conversation**: what do other people think about this information?
  - e.g. how do these options impact on the marketing, finance, and logistics team? Can they follow a similar strategy, or is it impossible for them to make the necessary changes?

It is clear to see that knowledge always requires human interaction. It does not simply exist; it needs to be understood, interpreted and reinterpreted. Knowledge therefore also includes practical experiences, insights, and know-how, which can be produced when people interact and communicate with each other. It constantly changes and is very context- and process-specific. This makes it difficult to effectively 'manage' knowledge, as it cannot simply be written down, stored or shared easily (Nonaka & Takeuchi, 1995; Nonaka & Konno, 1998; Bergeron, 2003; Choo, 2006). Dalkir (2017: 2) summarised a few more key characteristics of knowledge as follows:

- Using knowledge does not consume it.
- Transferring knowledge does not result in losing it.
- Knowledge is abundant, but the ability to use it is scarce.
- Much of an organization's valuable knowledge walks out the door at the end of the day.

Explicit and tacit knowledge

Knowledge can further be categorised into two different types: explicit knowledge and tacit knowledge. This idea is based on Polanyi's (1966, reprinted 1983: 4) book, *The Tacit Dimension*, where he argues that, "we can know more than we can tell." In other words, the two forms of knowledge can be defined along a continuum: explicit knowledge on one side of the continuum, can be expressed in words and numbers. It can be shared and communicated easily through, for example, formulas, principles and procedures. It can also be documented to some extent. On the other end of the continuum, tacit knowledge is very difficult to capture and share with other people. It is part of people's actions, experiences and beliefs, and therefore very individual, relational and context-specific (see for example, Nonaka & Takeuchi, 1995; Argote & Ingram, 2000; Von Krogh, 2002; Nonaka & Von Krogh, 2009; Dalkir, 2017). Knowing how to ride a bicycle, for example, would be classified as tacit knowledge: we can learn how to ride through practice and experience, but we cannot normally explain the actual process of how riding a bicycle works, unless we know anatomy, physiology and all the muscles involved in the process, as well as the mechanical details of the bicycle itself. In other words, tacit knowledge cannot be fully codified and made explicit, it may even be subconscious (Hislop et al., 2018). But we can learn it over time through observation, imitation, and practice. Table 1.1 summarises some of the key properties of both explicit and tacit knowledge.

It is important to note that tacit knowledge can further be distinguished along two different dimensions: (1) the technical dimension of tacit knowledge, which includes for example, skills, crafts, and know-how; and (2) the cognitive dimension of tacit knowledge, including beliefs, values, ideals, and mental models, which are part of who we are and which we often take for granted (Nonaka & Konno, 1998). Again, both these dimensions of tacit knowledge are difficult to capture, as well as difficult to share with others. In that sense, they can be a source of competitive advantage (competitors cannot easily copy them). At the same time, they can also be difficult to share within the organisation itself and to be made sense of and understood by all employees. Knowledge is therefore sometimes characterised as 'sticky' (Pervaiz et al., 2001: 11), because it is usually embedded in and stays in people's heads.

Table 1.1: Properties of explicit and tacit knowledge

Explicit knowledge	Tacit knowledge
Words, numbers, formulas	Experiences, beliefs
Can easily be accessed, (re-)produced, disseminated and (re-)applied throughout the organisation	Mainly used to adapt, and to deal effectively and efficiently with new and challenging situations
Employees can be taught and trained	Difficult to teach; a lot of it is expertise, know-how and know-why
Can be organised and systematised (e.g. to translate a vision into a mission statement and operational strategies)	Difficult to organise, but important for collaboration, for building a shared vision and understanding, as well as organisational culture
Is transferred via products, services and documented processes	Can to some extent be transferred through coaching and mentoring on a one-to-one, face-to-face and context-specific basis

Individual, group and organisational knowledge

The third important distinction in knowledge management is along the lines of where knowledge resides. This can be at the individual level, the group or team level, or the organisational level. First, of course, employees in any organisation hold a lot of individual knowledge; knowledge they have acquired over time, in different contexts and based on their individual background, expertise and experience. It helps them do their jobs, make decisions and be effective in their day-to-day tasks. It also provides opportunities to contribute and challenge certain decisions made by others (Alavi & Leidner, 2001; Dalkir, 2017).

Second, when individuals share their knowledge with others, group or team knowledge can be developed. This is ideally more than the sum of its parts. Work teams create knowledge around certain processes and practices they engage in on a regular basis, and they collaborate, network, and develop professional skills together. At a group level, a common language is therefore quite often developed from this knowledge (Dalkir, 2017). For example, in an event organisation, the marketing team holds specific group knowledge around designing posters, writing press releases, and promoting the event on social media. Some of the language they develop around these practices cannot easily be transferred onto other teams and does not usually mean a lot to, for example, the logistics or human resource team. It is very valuable to the marketing group though and creates a shared understanding among members of the group.

A danger that comes with creating group knowledge, is that of groupthink, especially if trust within the group and the group's cohesiveness are high. In other words, if all group members acquire similar kinds of knowledge (organisational knowledge, knowledge around specific processes and practices, 'the way things are done' within the group), they will inevitably stop challenging and questioning certain decisions and therefore get stuck and not create any new knowledge anymore (Zheng, 2009). They will stop to critically assess, analyse and evaluate their practices and decisions. Hence, while a shared understanding and shared language are beneficial to groups to some extent, too much of it can create barriers to the creation of new knowledge in the long-run and should be avoided.

At the group level, relationships and ways of working together are regarded as crucial. This becomes even more important at the third level – the organisational level. The organisation as a whole should hold valuable knowledge about its history, vision, values and beliefs, which help individuals and groups understand how they can contribute to the organisation's overall knowledge (Brown & Duguid, 1998, 2001; Ardichvili et al., 2003). Knowledge at the organisational level is also important for best practices to be shared, for solving problems quickly and efficiently, and for creating organisational memory. Ideally, the organisation as a whole constantly improves and innovates through the knowledge contributions of its individual and group/team members – a process commonly referred to as 'organisational learning' (Argyris & Schoen, 1978; Senge, 2006).

Knowledge management and organisational learning are therefore closely related concepts and they reinforce each other. Argyris and Schoen (1978) first introduced the term organisational learning and argued that it is a continuous process that organisations go through. Importantly, organisational learning cannot be achieved without individual learning, but individual learning alone is not enough for the organisation as a whole to also learn; there need to be other processes and practices in place in order for this organisational learning to happen. According to Argyris and Schoen (1978), two types of organisational learning can be distinguished: single-loop learning (also called adaptive learning) occurs when employees detect a match or mismatch and they then change their actions in order to correct it. This would be the case, for example, in routine and repetitive work practices, and mainly focuses on coping with smaller issues, problem-solving and incremental improvements, without actually questioning any existing ways of doing things and any existing ways of learning. Double-loop learning (or generative learning), the second type of organisational learning, might be necessary for more complex issues, and should be the main focus for most organisations. Rather than simply changing the actions to correct a problem or mismatch, employees would examine and alter the governing variables first, and therefore go through not just one, but two loops of learning. It is about continuous feedback and ongoing examination of how things are done in the organisation, and hence the development of new knowledge along the way. Double-loop learning is essential in creating new knowledge and effectively using it in all functions of the organisation's value chain. It can therefore develop into a source of competitive advantage or core competence. In other words, single-loop learning is about being adaptable, while double-loop learning is about having adaptability (Pervaiz et al., 2001). Both single- and double-loop learning, of course, require knowledge and expertise in order for employees to be able to detect issues, change them, and learn from the process. This in turn produces new knowledge for the individuals and groups involved, as well as for the organisation as a whole, which ideally will be able to develop an institutional or corporate memory consisting of archived experiences, routines, processes, actions and decisions. Learning and knowledge are thus in a constant cycle of mutually reinforcing each other (Argote & Ingram, 2000; Argote et al., 2000; Pervaiz et al., 2001; Senge, 2006; Hislop et al., 2018).

Knowledge management

Based on these different dimensions of data, information, and knowledge; explicit and tacit knowledge; and the three different levels of knowledge (individual, group, organisational), the successful management of knowledge has been defined in many different ways. Early definitions include, for example:

- The overall purpose of knowledge management is, "to maximize the enterprise's knowledge-related effectiveness and returns from its knowledge assets and to renew them constantly. KM is to understand, focus on, and manage systematic, explicit, and deliberate knowledge building, renewal, and application – that is, manage effective knowledge processes (EKP)" (Wiig, 1997: 2);

- Knowledge management is a "discipline that promotes an integrated approach to identifying, capturing, evaluating, retrieving, and sharing all of an enterprise's information assets. These assets may include databases, documents, policies, procedures, and previously un-captured expertise and experience in individual workers" (Duhon, 1998);

- Knowledge management is "about harnessing the intellectual and social capital of individuals in order to improve organizational learning capabilities" (Swan et al., 1999: 264);

- Knowledge management is, "a deliberate, systemic business optimization strategy that selects, distils, stores, organizes, packages and communicates information essential to the business of a company in a manner that improves employee performance and corporate competitiveness" (Bergeron, 2003: 8-9);

- Knowledge management is, "the deliberate and systemic coordination of an organization's people, technology, processes, and organizational structure in order to add value through reuse and innovation. This is achieved through the promotion of creating, sharing, and applying knowledge as well as through the feeding of valuable lessons learned and best practices into corporate memory in order to foster continued organizational learning" (Dalkir, 2017: 4).

Hislop et al. (2018) further highlight that knowledge management also includes other factors in organisations, such as the organisational structure and culture, its human resource practices and processes, as well as leadership styles. Any evaluation of organisational processes and practices also forms part of this, whereby information and knowledge can be gathered and their value for the organisation as a whole determined

(Getz, 2018). Knowledge management is therefore about effectively managing the knowledge processes and the knowledge work within an organisation, rather than managing knowledge itself. Any organisation needs to understand, analyse and make effective use of its various kinds of knowledge, residing in documents, processes, procedures, as well as in people. It has further been noted that, "[i]n any organisation, knowledge management takes place. How formalised or non-formalised it is, is debatable" (Du Plessis, 2006: 139).

For the purpose of this book and for event organisations in particular, knowledge management can therefore be defined as:

> *The effective use of organisational systems, processes and practices which allow both explicit and tacit knowledge to be created, identified, acquired, utilised, shared and stored, in order for the organisation to produce a successful event experience and to sustain a competitive advantage over time.*

A number of different knowledge management frameworks and models have been developed over time and these will be discussed further in Chapter 3. At this stage, however, Figure 1.2, developed by Stadler (2019: 154), brings together the different elements covered in this chapter:

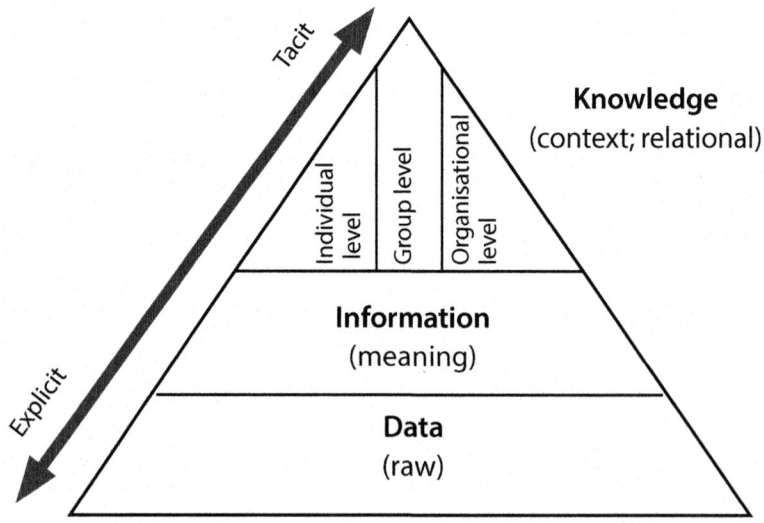

Figure 1.2: Dimensions and levels of knowledge. Source: Stadler (2019: 154), reproduced with permission of Edward Elgar Publishing Limited.

Lastly, the broader knowledge management literature distinguishes between three generations of knowledge management:

1. First generation knowledge management uses knowledge storage and retrieval systems to enhance organisational processes. It is sometimes also called the 'engineering perspective' (Swan, 2004), 'content perspective' (Hayes & Walsham, 2003) or 'object perspective' (Carlsson et al., 1996), simply because knowledge is classified as an object that can easily be stored and transferred. Technology is the answer to everything at this stage, but when looking into it a bit more, this generation of knowledge management really only covers Information Management. It will be covered in some sections of this book, but only insofar as it underpins and contributes to other (second and third generation) knowledge management processes and practices. In this sense, the book acknowledges that ICT and knowledge management are inseparable, but ICT alone is not enough for knowledge processes and practices to be effectively managed, and people and other factors also need to be taken into account.

2. Second generation knowledge management links elements of knowledge management to other organisational factors, such as processes, strategies, structures and organisational cultures. It is based on the above discussed notion that knowledge is *more* than just information.

3. Third generation knowledge management takes these ideas a step further and introduces knowledge practice theory and relational knowledge theory. This is where knowledge is seen to be situated in people's heads and therefore an emphasis is placed on human interactions, knowledge as being 'practised' by employees and therefore constantly changing and evolving. Swan (2004) calls this the 'community perspective', while Hayes and Walsham (2003) refer to it as the 'relational perspective'. It also includes broader questions around power and knowledge and alternative approaches to knowledge management, such as the Appreciative Sharing of Knowledge.

Study and discussion questions

- Provide a question and answer for each of the Learning Objectives. This should be done for all the chapters.
- Define explicit and tacit knowledge and provide examples for both these types of knowledge from an event you have organised or volunteered at.
- Discuss the difference between knowledge management and organisational learning. Which one do you think is more important for event organisations and why?
- Based on what you have learned so far, come up with your own definition of 'knowledge' and 'knowledge management'.

Recommended additional readings

Alavi, M., & Leidner, D. (2001). Review: knowledge management and knowledge management systems: conceptual foundations and research issues. *MIS Quarterly*, 25(1), 107-136.

Polanyi, M. (1966, reprinted 1983). *The Tacit Dimension*. Gloucester, MA: Peter Smith.

References

Alavi, M., & Leidner, D. (2001). Review: Knowledge management and knowledge management systems: Conceptual foundations and research issues. *MIS Quarterly*, 25(1), 107-136.

Ardichvili, A., Page, V., & Wentling, T. (2003). Motivation and barriers to participation in virtual knowledge-sharing communities of practice. *Journal of Knowledge Management*, 7(1), 64-77.

Argote, L., & Ingram, P. (2000). Knowledge transfer: A basis for competitive advantage in firms. *Organizational Behavior and Human Decision Processes*, 82(1), 150–169.

Argote, L., Ingram, P., Levine, J. M., & Moreland, R. L. (2000). Knowledge transfer in organizations: Learning from the experience of others. *Organizational Behavior and Human Decision Processes*, 82(1), 1-8.

Argyris, C., & Schoen, D. (1978). *Organizational Learning: A Theory of Action Perspective*. London: Addison-Wesley Publishing Company.

Bergeron, B. (2003). *Essentials of Knowledge Management*. Hoboken NJ: John Wiley & Sons.

Brown, J. S., & Duguid, P. (1991). Organizational learning and communities-of-practice: Toward a unified view of working, learning, and innovation. *Organization Science*, 2(1), 40-57.

Brown, J. S., & Duguid, P. (2001). Knowledge and organization: A social-practice perspective. *Organization Science*, 12(2), 198-213.

Carlsson, S. A., El Sawy, O., Eriksson, I. V., & Raven, A. (1996). *Gaining Competitive Advantage Through Shared Knowledge Creation: In Search of a new design theory for strategic information systems.* Paper presented at the ECIS, Lisbon, Portugal.

Choo, C. W. (2006). *The Knowing Organization – how organizations use information to construct meaning, create knowledge, and make decisions* (2nd ed.). Oxford: Oxford University Press.

Dalkir, K. (2017). *Knowledge Management in Theory and Practice* (3rd ed.). Cambridge, MA: MIT press.

Davenport, T. H., & Prusak, L. (1998). *Working Knowledge: How organizations manage what they know*. Harvard Business Press.

Du Plessis, M. (2006). *The Impact of Organisational Culture on Knowledge Management*. Oxford: Chandos Publishing.

Duhon, B. (1998). It's all in our heads. *Inform*, 12(8), 8-13.

Getz, D. (2018). *Event Evaluation*. Oxford: Goodfellow Publishers.

Gorelick, C., Milton, N., & April, K. (2004). *Performance through Learning - Knowledge Management in Practice*. Amsterdam: Elsevier Butterworth-Heinemann.

Hayes, N., & Walsham, G. (2003). Knowledge sharing and ICTs: A relational perspective. In M. Easterby-Smith & M. A. Lyles (Eds.), *The Blackwell Handbook of Organizational Learning and Knowledge Management* (pp.54-77). Blackwell Publishing.

Hislop, D., Bosua, R., & Helms, R. (2018). *Knowledge Management in Organizations - A Critical Introduction* (4th ed.). Oxford: Oxford University Press.

Nonaka, I., & Konno, N. (1998). The concept of 'Ba': Building a foundation for knowledge creation. *California Management Review*, 40(3), 40-54.

Nonaka, I., & Takeuchi, H. (1995). *The Knowledge Creating Company – How Japanese companies create the dynamics of innovation*. New York: Oxford University Press.

Nonaka, I., & Von Krogh, G. (2009). Tacit knowledge and knowledge conversion: controversy and advancement in organizational knowledge creation theory. *Organization Science*, 20(3), 635-652.

Pervaiz, K. A., Lim, K. K., & Loh, A. Y. E. (2001). *Learning through Knowledge Management*. Oxford: Butterworth Heinemann.

Polanyi, M. (1966, reprinted 1983). *The Tacit Dimension*. Gloucester, MA: Peter Smith.

Senge, P. M. (2006). *The Fifth Discipline - The Art and Practice of the Learning Organization* (2nd ed.). London: Random House Business Books.

Stadler, R. (2019). Knowledge management in event and festival organisations: Challenges and future directions. In E. Lundberg, J. Armbrecht, & T. Andersson (Eds.), *A Research Agenda for Event Management* (pp. 154-169). Cheltenham: Edward Elgar.

Swan, J. (2004). Knowledge management in action? In C. Holsapple (Ed.), *Handbook on Knowledge Management - I. Knowledge Matters* (pp. 271-296). Berlin: Springer-Verlag.

Swan, J., Newell, S., Scarbrough, H., & Hislop, D. (1999). Knowledge management and innovation: networks and networking. *Journal of Knowledge Management*, 3(4), 262-275.

Von Krogh, G. (2002). The communal resource and information systems. *Journal of Strategic Information Systems*, 11(2), 85-107.

Wiig, K. M. (1997). Knowledge management: Where did it come from and where will it go? *Expert Systems with Applications*, 13(1), 1-14.

Zheng, W. (2009). The knowledge-inducing culture - an integrative framework of cultural enablers of knowledge management. *Journal of Information & Knowledge Management*, 8(03), 213-227.

2 Knowledge Management Challenges in Event Organisations

Learning objectives

- Understand different knowledge management challenges for event organisations.
- Be able to identify knowledge management challenges for project-based and 'pulsating' types of organisations.
- Understand how the large number of seasonal employees and volunteers in event organisations impacts upon knowledge management.
- Explore the different event specific types of knowledge (creative, operational, strategic).
- Learn the difference between knowledge hiding and knowledge hoarding.
- Be able to explain how the different knowledge management challenges impact upon reinventing the wheel.

Introduction

This chapter sets the scene for knowledge management in an events context. The events industry is a highly competitive industry, where many new events emerge, and unsuccessful ones disappear. Events management thus needs to be effective in order for the organisation to be successful, both in economic as well as in creative terms. The notion of success is thereby "(...) as much an inward-looking concept as an outward one" (Getz & Frisby, 1988: 23). Effective knowledge management can help event organisations stay innovative and competitive in the long term. This, however, comes with a few challenges.

Allen et al. (2011) and Bowdin et al. (2012) provide an extensive overview of the managerial process of organising special events. An operational focus includes strategic considerations, marketing, financing, human resource management, logistics, legal issues, and risk management. Allen et al. (2011: 495) also identified the issue of knowledge management as part of the post-event evaluation, where,

> "[t]he staging of major events and conferences has now become so complex that event managers and organising bodies cannot afford to start from scratch in the planning of events. They must start from what has been learnt from the previous staging and history of the event and build on this (...). [T]his process of the transfer of knowledge takes place partly through the documentation of the event and partly through the skills and experience of key event personnel, who become highly sought after because of their successful track record in organising events."

Many event organisations, however, fail to document and share the acquired knowledge and end up reinventing the wheel each event season. In the wider management literature, an organisation's basic characteristics, such as its size (i.e. number of employees), whether it is a private organisation or a public sector organisation, whether it is geographically dispersed (e.g. multinational corporations), and whether it is run very formally or more informally (such as many small- and medium-sized enterprises), all play an important role in determining the best and most effective way of managing knowledge processes and practices (Hislop et al., 2018). There are, however, several additional knowledge management challenges specifically relevant to the field of events, which will be discussed in this chapter. These are: the 'pulsating' nature of events; lack of time and resources to create, share and document knowledge; large numbers of volunteers; knowledge leakage, knowledge hiding and knowledge hoarding; as well as the team moving on once the event is over, and knowledge therefore being lost. Knowledge management in project-based organisations will also briefly be discussed in this chapter, as there are a number of similarities between these types of organisations and events.

The 'pulsating' nature of events

With one-off events, of course, there is no previously accrued organisational knowledge available for staff members to draw on. They need to quickly and efficiently engage in the process of creating new knowledge, which can be a massive challenge. But even with recurring events, such as annual events, staff turnover is high: when the event is approaching, the organisation expands rapidly, new seasonal staff members and volunteers join and need to quickly gain an understanding of key processes, practices and procedures within the organisation. They need to acquire knowledge, share existing knowledge with others, and together create new knowledge. As soon as the event is over, however, the organisation then contracts again, and staff members move on to other jobs (Carlsson-Wall et al., 2017; Clayton, 2020).

Toffler (1990) originally defined these types of organisations as 'pulsating' organisations. Figure 2.1 below depicts the number of people involved over time in a typical 'pulsating' event organisation running, for example, one event per year.

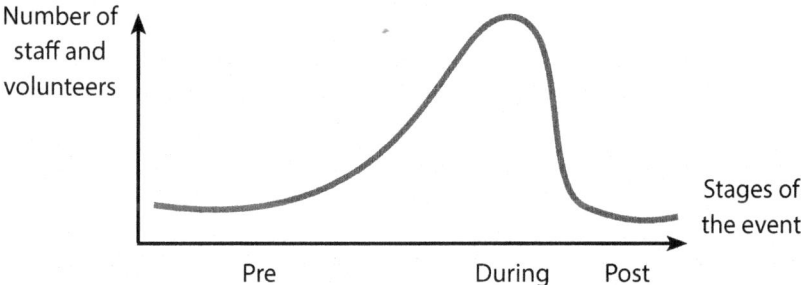

Figure 2.1: 'Pulsating' event organisations

Hanlon and Cuskelly (2002) later applied the same idea to mega sports events and argued that event organisations share similar characteristics to these 'pulsating' organisations. They specifically looked at the induction process for new staff members in these types of organisations, where there is no stable workforce in place and therefore no established relationships between staff members. Induction tends to be run at a group, rather than individual, level in 'pulsating' event organisations because of the massive influx of people over a short period of time. Hanlon and Cuskelly (2002) highlight the importance of induction manuals to facilitate the process and recommend running induction sessions at all different levels of the organisation and for both full-time and seasonal staff members. These are effective ways of sharing explicit knowledge, but the more complicated process of creating and sharing tacit knowledge also needs to be considered.

Furthermore, there are two types of 'pulsating' event organisations: 'single pulse' event organisations, which are one-off events only and which grow, decline and dismantle once within the same location; whereas 'repeat pulse' event organisations are organisations, which go through the entire cycle perhaps annually or biennially (Lockstone-Binney et al., 2020). Particularly for this second type of repeat pulse organisations, knowledge management becomes crucial in terms of building a long-term organisational memory so as not to start again from scratch each year. In terms of 'pulsating' organisations and their knowledge management processes, not surprisingly, the following more specific challenges therefore need to be overcome:

1. There are no existing trust relationships between staff members and temporary staff/volunteers, which means individual staff members might end up hoarding knowledge rather than openly sharing it with others;

2. New staff members might bring a lot of knowledge with them, but find it difficult to apply this knowledge in an unfamiliar context and with new team members, especially if a lot of their knowledge is tacit in nature;

3. Knowledge about the event itself and how things are done (a shared vision and understanding) needs to be shared with new staff members quickly and efficiently, but not all of it will be readily available or even relevant to them, again, especially if it is in tacit form and therefore takes a long time to acquire;

4. A lot of new knowledge will undoubtedly be created along the way, and especially during the event itself. But time and resources are limited to share and document this newly created knowledge for future years; and

5. A lot of knowledge will be lost post-event when staff members run out of time to document their knowledge before leaving the organisation and when they move on to their next job.

In addition to this, the (usually) large number of volunteers poses another knowledge management challenge to event organisations, as well as the issue of bringing creative, operational and strategic knowledge together, as discussed further below. An overview of all these challenges can be found in Figure 2.2, and each of them will be further discussed in the following sections.

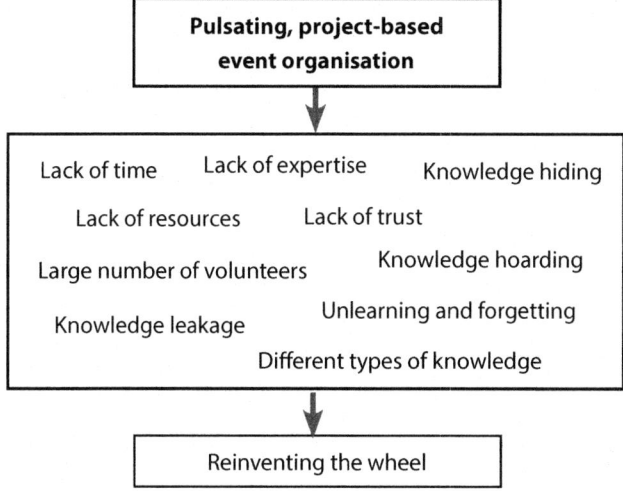

Figure 2.2: Knowledge management challenges in 'pulsating', project based event organisations

Knowledge management in project-based organisations

It is worth noting that project-based organisations face similar knowledge management challenges to event organisations, and these have been well researched and documented in the wider knowledge management literature (for recent studies, see for example, Duffield & Whitty, 2015; Winkler & Mandl, 2015; Ghobadi & Mathiassen, 2016; Pereira & Gonçalves, 2017). It should be stressed, however, that most studies on knowledge management in project-based organisations are limited to large construction and manufacturing firms, communications and the media, and consulting firms (Kodama, 2007), and most of them focus on new product development projects (Bresnen et al., 2005). However, the overall idea of event management is the management of an *experience*, and knowledge management in event organisations is thus even more complicated than within a factory or firm. There are some insights from the film industry though, which has received particular attention here and will be summarised below.

It is oftentimes argued that within projects, knowledge is created and shared between experts, and projects therefore provide the basis for the creation of new knowledge and for innovation. However, at the same time, the short-term nature of projects and the attention that has to be given to schedules and deadlines, make it difficult to build a common ground for sharing knowledge among project team members, let alone the necessary commitment and motivation. Furthermore, the lessons learned from a project are difficult to document or transfer to other projects or the organisation as a whole. When the project is over, a lot of knowledge is lost (Lewis, 1998; Scarbrough et al., 2004; Liebowitz, 2005; Lindkvist, 2005). In this regard, projects share similar characteristics with events.

In a comparative analysis of two case studies of construction projects, for example, Scarbrough et al. (2004) found that projects have a high potential for the generation of learning. However, this kind of project-learning is very much shaped by the ongoing learning activities of the organisation as a whole. Furthermore, it is situation and context specific, and bound to the very project. The exploitation of such learning for the wider organisation is thus limited. In a later study, Swan et al. (2010) dealt with the question of how the organisational context influences learning from projects. They found that organisations, which are centred on projects and where project management is highly developed, are successful in accumulating experience and learning. Organisations with only occasional and very varied types of projects, on the other hand, are less likely to succeed in knowledge sharing and learning. This is an interesting point for event organisations to consider, because events are, on the one hand, infrequent and thus fall in the latter category of 'projects' suggested by Swan et al. On the other hand, the process and the event cycle are similar each year and usually well developed in terms of organising and staging the event. Therefore, they can accumulate knowledge, experience and learning over time.

DeFillippi and Arthur (1998) studied the creative project-based work of the film making industry, and thus provide some ideas which are more closely related to events. They highlight that the film industry is characterised by temporary employment and the collaboration of a variety of specialists working together. Once the film is released, the producing company dissolves; only sometimes do individuals work together again in subsequent film projects. The authors point out that both human capital and social capital are crucial in building a film crew and producing a film. That is, the industry highly depends on experts who know their field (both artistic experts, as for example script writers, set designers, art directors, or sound mixers; as well as commercial experts associated with the project's financial success, such as producer, production manager, or production accountant), and also the soft infrastructure of relationships and networks. DeFillippi (2015) later identified three key parameters for effectively managing project-based organisations in the creative industries, such as theatre productions, music, films, video games, advertising companies, or software development: roles, relationships, and routines. All three are crucial for the generation of new knowledge, as well as the adaptation of this knowledge to different work tasks. Learning in these types of organisations takes place in episodes for both the industry as a whole, as well as for individual participants (DeFillippi & Arthur, 1998). Some collective memory is built, for example, through each film project and things that worked or did not work. At the same time, participants and crew members build their own memory and experience in the field. However, "what is distinctive here is that there is no place for 'organizational memory' as conventionally presented in strategic management theory" (DeFillippi & Arthur, 1998: 136).

Events similarly provide the opportunity for building collective memory for the entire industry (for example through operational processes, marketing, or logistics), as well as for individual staff members. They also rely on expert knowledge and people with the relevant skills and expertise, as well as on relationships and networks; and they also follow a similar cycle of episodic learning. However, what distinguishes them from the film industry is that each event organisation *is* able to build an organisational memory and can learn over time even though individual members might leave after the event is over. The core group, or permanent staff, ideally, stay with the organisation and can thus help to build this organisational memory over time, especially within the 'repeat pulse' type of event organisations as defined above.

Lack of time and resources

For event organisations, lack of time and resources is a particular challenge that can have significant impacts on knowledge management (see for example, Muskat & Deery, 2017). Having an influx of people at different stages during the event life cycle means that knowledge needs to be shared with them quickly and efficiently. However, if they all have different needs, attitudes and previous knowledge and expertise, it can be nearly impossible to effectively manage the induction, training, and on-the-job learning processes required for them to be at their optimum level of performance in time for the event to kick off. Due to time pressures in the weeks and days leading up to the event, induction and training sessions often end up being run inconsistently or haphazardly. Or, in some cases, it might be possible to teach seasonal staff members the required skills to do their jobs, but they still lack an understanding of how the organisation as a whole works and how their roles are related to the organisation as a whole (Hanlon & Cuskelly, 2002). Event organisations usually lack the necessary support structures, routines and resources for knowledge management practices to be built into induction and training sessions.

Furthermore, once the event is in full swing, all staff members (permanent and seasonal staff, as well as volunteers) are busy performing their day-to-day tasks. Unquestionably, they will create a lot of new knowledge together, but will not have the time to effectively share and document it and hence make it explicit. Particularly at the team or group level, at this stage, staff members usually create a lot of shared routines, a shared language and understanding of 'how things are done', and hence a lot of tacit knowledge. This, however, is difficult to integrate into the organisation as a whole and for future years, due to lack of time and resources. Ragsdell et al. (2013) argue that seasonal staff members might not even be aware of the value of doing so for the event organisation as a whole. Because time is limited, the willingness and opportunities to integrate new knowledge into the wider organisation are also limited. This further affects the establishment of deeper relationships in event organisations, where staff members might think they only need to know enough to do their day-to-day tasks, and do not need to share this knowledge with others. After all, at busy times and with limited resources available, they can only keep themselves up-to-date and informed in certain areas, knowing everything about all other areas of the organisation would simply lead to information overload.

One specific issue here, is that of time pressure and dealing with deadlines. In his paper about time and timing in projects, Rämö (2002) applies the idea of chronos-time (the clock-time, the exact quantification of time) and kairos-time (doing the right things at the right moment, regardless of clock-time) to project management. He argues that both types of time are an issue for project management and proposes the idea that chronos-time is crucial for efficiency, whereas kairos-time moments are essential for effectiveness. This is also true for event organisations: on the one hand, they must operate 'on time' and 'by the clock', and deliver the product – that is, the event – on time. However, they must also "pay attention to the more creative aspects of time, when a feeling for the right moment to act can result in unfolding new and bold ideas. Anything else is a waste of time" (Rämö, 2002: 573). Emery and Radu (2008) apply the same idea to the management of major sport events and argue that in the events industry decisions are often made within seconds or minutes, and they are on the one hand based on experience and a common understanding, on the other, intuition and a feeling for the right thing to do is also often necessary. Events and festivals definitely face the challenge of only having one opportunity to get it right.

Lastly, once the event is over, post-event evaluations and reviews are often neglected or overlooked as staff members and volunteers are eager to move on to their next project, event or other jobs. The opportunity to reflect on lessons learned and store them in the organisation's memory is therefore lost due to time constraints and lack of resources at this stage. This is particularly the case when there is no specific succession plan in place and no specified routines and practices, such as exit interviews (Maaninen-Olsson et al., 2008; Ragsdell et al., 2013).

Lack of expertise and trust among the team

Staffing, recruiting and retaining the right employees in an event organisation is of course very important, and has received considerable attention in the wider event management literature (see for example, Van der Wagen, 2007; Beaven et al., 2009; Deery, 2009; Smith & Lockstone, 2009; Allen et al., 2011; Lockstone-Binney et al., 2020). A particular challenge for event organisers is having to rely on people's previous experience rather than on training, because as discussed above, time is limited for conducting training sessions with everyone. Event organisers also have to rely on the expertise of other stakeholders, such as contractors, suppliers, and partners, to carry out certain elements and functions of the event (e.g. catering, staging, security). If employees and/or contractors lack the necessary skills and expertise to do these tasks, this can sometimes be a reason for events and festivals to fail (Getz, 2002; Hanlon & Cuskelly, 2002). Training and on-the-job learning therefore need to be very targeted, effective and efficient. They form key processes for employees and other stakeholders to acquire knowledge, share their existing knowledge with others and together create new knowledge, and are hence important for effective knowledge management within the organisation overall.

A further challenge related to this is lack of trust among the team. Even if each individual employee and contractor does have the necessary skills and knowledge to carry out their tasks, in many cases they will come together for the very first time to work on the event and will lack the trust relationships that are necessary for effective teamwork. Hinds and Pfeffer (2003: 15) highlight that even if employees have the motivation to share their knowledge with others, it will not happen unless they trust each other as well as the organisation as a whole: "Being motivated to share what you know with others requires trust – not only trusting those others (something that is diminished with competition) but also trusting the larger institution within which the sharing of expertise is occurring." However, it can be very difficult to build trust within a short period of time, and in an events context, this sometimes means within a week or even just a few days before the event takes place.

Trust is therefore a key factor to consider not just in terms of knowledge management, but for any group and organisation to work together effectively: "An essential aspect of group effectiveness is developing and maintaining a high level of trust among group members. The more members trust each other, the more effectively they will work together" (Johnson & Johnson, 2003: 124). If event organisers want their employees to cooperate with each other, "the crucial elements of trust are openness and sharing on the one hand and acceptance, support, and cooperative intentions on the other" (ibid: 128). All of these are 'soft' factors of knowledge management and will be discussed further in Chapter 6.

It is worth noting that in a study by Ragsdell and Jepson (2014) on CAMRA festivals and festival volunteers in particular, different types and forms of trust were identified as both knowledge sharing enablers on the one hand, as well as barriers to sharing knowledge on the other. For example, trust in the organisation itself and its processes, trust in the quality of the knowledge that is being shared, as well as trust in the abilities of the volunteers, together created a positive environment for knowledge to be shared. Yet at the same time, for some volunteers too high a level of trust was actually seen as a barrier to sharing knowledge. This was mainly because they started to feel complacent, and in some occasions identified groupthink to set in, which led them to think that it was then unnecessary to create new knowledge together. Finding the right balance and level of trust is therefore crucial, but very difficult to achieve in practice.

Large number of volunteers

Human resource management in events has received considerable attention over the last few years (Van der Wagen & White, 2014), and volunteer management in particular is an ongoing topic for investigation, most notably in sports and mega events (see for example, Smith & Lockstone, 2009; Smith et al., 2019; Lockstone-Binney et al., 2020). The field of event management is becoming more and more professional (Mair, 2009), but it is commonly acknowledged that while a good educational background and knowledge and skills are important to have before entering the industry, there is also a lot of on-the-job training and learning necessary in order to apply this in practice and to further develop the practical skills (Arcodia, 2009; Junek et al., 2009). This is particularly the case with volunteers. Current literature on event volunteering, however, mainly emphasises individual volunteers' motivations and experiences, rather than taking an organisational or institutional approach (Smith et al., 2019), where volunteers' knowledge creation, transfer and documentation practices can potentially benefit the organisation as a whole.

For most events volunteers play a major role: there are usually a large number of people who are passionate about the event and who enjoy helping to deliver the event experience without receiving any monetary compensation. Volunteers, however, need as much management and motivation as paid staff, otherwise they will burn out or lose their passion and enthusiasm for the job (Smith et al., 2019). Working at an event is an intense experience and the workloads are oftentimes too high. In addition to this, volunteers can be the only direct encounter that attendees have with a representative of the event team; be it in the car park, at the entrance, or the information booth. Volunteers, as well as those staff members that are likely to personally deal with attendees, need to be trained in service relationships. If volunteers lack people skills and knowledge about the event itself, it will negatively affect the image of the event. Unsatisfied attendees in turn can easily ruin an event's reputation (Drummond & Anderson, 2004).

As mentioned above, recruiting the right people in the first place is essential in order to have all areas and skills covered. The process of recruitment and selection is about finding the 'right' people for the job. It is also an opportunity for applicants to show that they are interested in the position and that they have the required skills and knowledge to do the job. With volunteers, however, this is often overlooked. When selecting volunteers, emphasis is usually put on their past performance and experience, background knowledge, reasons for applying, and understanding of the event and position. Quite often, however, event organisations are desperate for any additional help available and therefore anybody applying to be a volunteer will be given a chance to help out at the event, no matter their level of expertise and knowledge. Furthermore, induction sessions tend to be run at a very generic level, rather than specific training providing new volunteers with the necessary information and knowledge in order to perform their individual roles. Ad hoc on-the-job training with other staff, experts, or in teams seems to be common in event organisations, which can be a very effective method; however, existing staff sometimes do not know everything or cannot put certain things in the right context so that newcomers understand them (Allen et al., 2011).

Volunteers also want to feel needed and useful to the organisation. However, for event volunteers this can be a challenge, especially at the beginning when they feel unprepared and not quite 'ready' to do the job. It is therefore important to enhance those conditions that enable effective knowledge sharing, while at the same time remove any potential barriers to knowledge sharing. And lastly, it is also important to retain volunteers and ensure they come back each year, so that the organisation can benefit from having volunteers in place who already know what they are doing or have the expertise and skills to mentor others on the job. Succession planning is key to the long-term success of an event organisation and this should, of course, include volunteers (Smith et al., 2019), particularly when it comes to managing their knowledge over time.

Creative, operational and strategic knowledge

Managing an event from start to finish is a highly creative process where staff members and volunteers are aiming to create an *experience* rather than a product or service (Berridge, 2007). This creative and very event specific knowledge underpins all other operational processes, such as marketing, sponsorship, logistics, human resource management, and even risk management and security. Staff members need to have a shared understanding of how their creative ideas impact upon other functions and activities within the organisation (Stadler & Fullagar, 2016), yet oftentimes the creative knowledge acquired and shared along the way is perceived as being less valuable than the operational knowledge of running the event. This can easily lead to a conflict of ideas, restrictions and hence lack of innovation in the long run. It is particularly the case with music festivals where the creative director's and the managing director's ideas might clash, or where one of them simply ends up using their power to implement *their* vision and ideas – whether they are feasible for the organisation as a whole or not. It is then up to the team to make this work, even though they may not believe in the same creative vision and ideas themselves. Other staff members may at the same time feel like their voices are not heard and they will therefore end up not sharing their ideas and knowledge with others anymore. All knowledge management practices are after all embedded in an organisational culture and identity that can enhance or impede new ideas, knowledge creation and organisational learning (see Chapter 6). If employees do not share this common understanding and the organisational culture, they can easily lose motivation and potentially end up hiding or hoarding their knowledge, an issue that will be discussed further below.

Furthermore, while the creative and operational knowledge are of course important for the day-to-day tasks within the event organisation, it is equally important to have a long-term strategic plan for the organisation as a whole in place and to evaluate this along the way (Allen et al., 2011; Bowdin et al., 2012; Blackman et al., 2017; Getz, 2018, 2019). Both forward planning and short-term delivery are vital for the success of any event organisation (unless it is a one-off event). But while the senior management team (or in some cases, the board of directors) may well have a strategic vision and plan for the organisation, seasonal staff members and volunteers are usually not aware of these long-term plans and may therefore lack an understanding of how their day-to-day tasks contribute to the long-term success of the organisation overall. In terms of knowledge management, there is no consensus on how much of this strategic knowledge needs to be shared with everyone. Too little understanding can lead to staff members thinking 'so what?' and 'why should I share?', whereas too much sharing can lead to information overload, which is just as dangerous in an already stressful environment such as in an event organisation.

Drawing on the wider knowledge management literature, however, it is commonly argued that a specific knowledge management/knowledge sharing strategy should be part of any organisation's broader strategic plan (Earl, 2001; Heisig, 2009). Any event organisation should assess and evaluate its impact (Getz, 2019), which includes determining whether they have achieved the main aims and goals, and then building their strategic plan from there. This should, of course, align with the overall aims, mission and objectives of the organisation. Yet, this can be very difficult to formalise within a 'pulsating', project-based organisation, where there may not even be a common vision and mission, let alone specific business or project goals and objectives (Ragsdell et al., 2013). With these not in place, employees will most likely fail to see the importance of sharing their knowledge with others and potentially end up hiding or hoarding their knowledge.

The issue of unlearning, forgetting and knowledge leakage

Issues of unlearning and forgetting are – surprisingly – not normally addressed in the wider knowledge management literature. It is, however, the case that organisations sometimes need to deliberately give up or abandon some of the knowledge that has been created over time, whereas at other times, they accidentally forget certain work routines. Unlearning is therefore, "the ability to deliberately dispose of or discard obsolete, outdated, or unwanted knowledge." Forgetting, on the other hand, "refers to the loss of knowledge that is not necessarily intended or planned" (Hislop et al., 2018: 127). In other words, when we unlearn, we aim to replace some of the old knowledge with new knowledge that is more relevant to the task at hand, or more in line with industry trends. The events industry is constantly changing, and a certain amount of organisational unlearning is therefore necessary in order for the organisation to make room for new approaches to, for example, technology or sustainability. This is not necessarily a bad thing. Unlearning can be a difficult process though, as it requires a lot of reflection as well as consciously giving up processes and practices that might be taken for granted or are embedded in the organisation's culture and value system.

Forgetting, on the other hand, is the process of getting rid of some of the knowledge altogether. It is unintentional and can have serious effects on the level of competitiveness and productivity of an organisation. De Holan and Phillips (2004) identified several different types of forgetting, two of which can specifically be applied to event organisations: 1) memory loss: organisational memory loss typically occurs when knowledge is only used infrequently. With many event-related processes and routines only being carried out once a year (in the case of, for example, an annual event), it is quite common for an organisation to lose the ability to carry out these work routines if they are unused, informal or not codified/made explicit; and 2) failure to capture new knowledge: in some cases, organisations accidentally forget to capture new knowledge, if this knowledge has only been created at an individual or group/team level. This knowledge then does not become institutionalised and is hence forgotten. It is particularly the case in project-based organisations, such as event organisations, where knowledge created at the project team level is not shared with the wider organisation or is not effectively codified/made explicit before the team moves on to the next project.

Lastly, knowledge leakage is related to the issue of unlearning and forgetting, but slightly different in that it has been defined as "vital/competitive organizational knowledge that leaves a safe environment, moving into or being released into another environment where this knowledge can be used or applied in either the same way or in new competitive ways" (Hislop et al., 2018: 137). This can happen at all three levels – the individual, group and organisational level, and is also a common issue in event organisations, where many seasonal staff members, volunteers, contractors and suppliers work for more than one organisation at a time and can therefore easily, either intentionally or unintentionally, share confidential information with rival organisations. It is particularly the case when an employee might feel unhappy, disgruntled about something, unfairly treated or underpaid, and can lead to a loss of competitive advantage for the organisation.

The issue of knowledge hiding and knowledge hoarding

Knowledge acquired on-the-job is usually believed to belong to the organisation, not the individual. But individual employees might think it is their personal intellectual property and hence they might not want to share it with others. Ideally, organisations should therefore aim to build a knowledge sharing culture, where employees are encouraged to work together, share their knowledge and collaborate, and where they are rewarded for doing this. However, employees are usually rewarded for what they know (their individual knowledge) and not for what they share with others (Bilginoğlu, 2019). Naturally, they might therefore start withholding their knowledge, hide it or hoard it. Any organisation is constantly at risk of failing when employees decide to withhold their knowledge from others and from the organisation as a whole. Event organisations are even more at risk of this happening, because of their temporary, 'pulsating' nature, where employees might not feel that they need to share their knowledge with others, if they only work together for a short period of time. Specifically, there are two ways employees can withhold their knowledge: knowledge hiding and knowledge hoarding. Both can happen in an event organisation.

Knowledge hiding is intentional. It happens when one employee requests knowledge from another, but this knowledge is intentionally concealed and not shared with them. In other words, specifically requested knowledge is being kept secret from the person who requested it. Knowledge hiding is hence closely related to questions around power, which will be discussed further in Chapter 7. According to Das and Chakraborty (2018), knowledge hiding can occur in three different ways:

- *Evasive hiding* means either providing impractical or irrelevant knowledge to the person who requests it, or waiting to share the knowledge until it is no longer of use and then promising to help at a later time;
- *Playing dumb* is simply achieved by pretending not to know the answer to a problem or not having the relevant knowledge readily available; and
- *Rationalised hiding* refers to hiding knowledge from someone by saying it is confidential and can therefore not be shared with anyone.

Knowledge hoarding, on the other hand, is unintentional. It happens when an employee (unintentionally) retains or accumulates knowledge, which then may or may not be shared at a later stage. Employees may thereby have the best intentions about getting their jobs done, but they simply do not realise that their knowledge would also be of value to others (Das & Chakraborty, 2018). Knowledge hoarding usually happens when there is competition among employees or between functional areas in an organisation, where employees start to work in silos. It is also more common in organisations where expertise and know-how are regarded as very important, while mentoring and helping others is regarded as less important. Husted and Michailova (2002: 65-67) identified a number of specific reasons as to why employees might not want to share their knowledge with others and rather hoard it:

- *Potential loss of value and bargaining power*: much of one's knowledge is 'hard-won' and therefore very valuable to the individual. They will therefore want to protect this individual competitive advantage from others;

- *Reluctance to spend time on sharing knowledge*: sharing knowledge requires a lot of time and effort (especially if the knowledge is tacit in nature) and may be perceived as an extra burden in an already stressful work environment;
- *Fear of hosting 'knowledge parasites'*: if an employee has put a lot of hard work into acquiring their knowledge, they might not want others to then tap into this knowledge and acquire it with less effort;
- *Avoidance of exposure*: when knowledge is shared, others will assess the quality of this knowledge. Employees might therefore feel reluctant to share their knowledge because they fear others will perceive it as poor-quality knowledge;
- *Strategy against uncertainty*: if knowledge is shared with others, there is a danger that the receiver of the knowledge will misinterpret it, and this could lead to negative consequences to the person who shared their knowledge;
- *Hierarchy and power*: employees at a lower level might not want to share their knowledge and come across as more knowledgeable than their superiors. At the same time, managers might hoard knowledge in order to maintain their power and hierarchical status.

In other words, when employees hoard their knowledge, they usually fear that the more they share, the more they will give away their power, especially if their knowledge is deemed particularly valuable. They will always also have that knowledge available to themselves and "do not need to compete with others or chase around the organization for a particular piece of information" (Bilginoğlu, 2019: 63). However, the danger that comes with this, is if a knowledge hoarder leaves the organisation, they will take all of that knowledge with them and their knowledge will be lost because there is no memory of it within the wider organisation. This can then easily lead to the organisation reinventing the wheel, a very common issue in event organisations.

Reinventing the wheel

Due to the 'pulsating' nature of events, many event organisations – and in particular the 'repeat pulse' types of event organisations – fail to create organisational memory because they do not document and share valuable knowledge for the future. In other words, "[i]f knowledge is not shared then wasteful cycles of re-learning can occur and there could even be significant failures in an organisation" (Ragsdell et al., 2013: 1). It can therefore be said that a combination of some or all of the factors and challenges discussed throughout this chapter may lead to an organisation reinventing the wheel every event season. Large-scale events, for example, are prone to face knowledge management challenges with regards to their volunteer workforce, lack of trust, and hence issues of knowledge hiding and hoarding. Small-scale events, on the other hand, may lack the necessary expertise within the team, as well as the resources to invest in databases and other tools to store and document some of the explicit knowledge. They may also be unaware of the importance of creating strategic knowledge alongside the knowledge required for day-to-day operational tasks.

Either way, a combination of the 'pulsating', project-based nature of events and the specific knowledge management challenges they face, will have an impact on the long-term success of the organisation. This is particularly the case with annual or biennial events where even the permanent staff cannot remember everything and may 'forget' key practices from one year to the next, let alone effectively share them with seasonal staff members and volunteers. The challenges discussed in this chapter are hence associated with organisational forgetting or memory loss, which is likely to occur when there are no standard processes in place for knowledge to be made explicit and to be shared/stored, when there is a lack of time and resources to engage in knowledge sharing activities and when there is high staff turnover. It is this organisational forgetting that leads to reinventing the wheel, as past successes need to be relearned each time, while failures will be experienced over and over again, and resources will therefore be wasted.

Lastly, as previously highlighted, managing an event means creating an *experience*, rather than a product or service. By nature, most of the knowledge created and acquired along the way is therefore tacit, and hence difficult to share with others and document for future use. It is, however, important for event organisations to create an organisational memory over time. This can be achieved by having an organisational structure and culture in place that enhances the transfer of knowledge. Later chapters in this book will provide some specific ideas of how this can be accomplished, or at the very least how event organisations can attempt to avoid reinventing the wheel, despite all the challenges highlighted here.

Study and discussion questions

- Define 'pulsating' organisations and explain how this concept can be applied to event organisations. Why is lack of time a specific knowledge management challenge in events?
- Discuss human resource management practices that can help manage volunteers' knowledge more effectively and efficiently pre-, during and post-event.
- Discuss how you would deal with an employee who is hiding or hoarding their knowledge.
- Based on your practical experience, how can event organisers avoid reinventing the wheel? Propose at least five specific strategies to overcome this issue.

Recommended additional readings

Clayton, D. (2020). Knowledge management in events. In S. J. Page & J. Connell (Eds.), *The Routledge Handbook of Events* (2nd ed., pp. 442-456). London and New York: Routledge.

Hanlon, C., & Cuskelly, G. (2002). Pulsating major sport event organizations: A framework for inducting managerial personnel. *Event Management*, 7(4), 231-243.

Ragsdell, G., Espinet, E. O., & Norris, M. (2013). Knowledge management in the voluntary sector: a focus on sharing project know-how and expertise. *Knowledge Management Research & Practice*, 12, 351-361.

References

Allen, J., O'Toole, W., McDonnell, I., & Harris, R. (2011). *Festival and Special Event Management* (5th ed.). Milton, Qld.: John Wiley & Sons.

Arcodia, C. (2009). Event management employment in Australia: a nationwide investigation of labour trends in Australian event management. In T. Baum, M. Deery, C. Hanlon, L. Lockstone, & K. Smith (Eds.), *People and Work in Events and Conventions - A Research Perspective* (pp. 17-28). London: CABI.

Beaven, Z., George, S. S., & Wright, R. (2009). Employability in the cultural events sector: The role of specialist degree programmes. In T. Baum, M. Deery, C. Hanlon, L. Lockstone, & K. Smith (Eds.), *People and Work in Events and Conventions - A Research Perspective* (pp. 29-38). London: CABI.

Berridge, G. (2007). *Events Design and Experience*. London: Routledge.

Bilginoğlu, E. (2019). Knowledge hoarding: A literature review. *Management Science Letters*, 9(1), 61-72.

Blackman, D., Benson, A. M., & Dickson, T. J. (2017). Enabling event volunteer legacies: A knowledge management perspective. *Event Management*, 21(3), 233-250.

Bowdin, G., Allen, J., Harris, R., McDonnell, I., & O'Toole, W. (2012). *Events Management*, London: Routledge.

Bresnen, M., Edelman, L., Newell, S., Scarbrough, H., & Swan, J. (2005). A community perspective on managing knowledge in project environments. In P. E. D. Love, P. S. W. Fong, & Z. Irani (Eds.), *Management of Knowledge in Project Environments* (Ch. 5). Oxford: Butterworth-Heinemann.

Carlsson-Wall, M., Kraus, K., & Karlsson, L. (2017). Management control in pulsating organisations - A multiple case study of popular culture events. *Management Accounting Research*, 35, 20-34.

Clayton, D. (2020). Knowledge management in events. In S. J. Page & J. Connell (Eds.), *The Routledge Handbook of Events* (2nd ed., pp. 442-456). London and New York: Routledge.

Das, A. K., & Chakraborty, S. (2018). Knowledge withholding within an organization: the psychological resistance to knowledge sharing linking with territoriality. *Journal on Innovation and Sustainability*, 9(3), 94-108.

De Holan, P. M., & Phillips, N. (2004). Organizational forgetting as strategy. *Strategic Organization*, 2(4), 423-433.

Deery, M. (2009). Employee retention strategies for event management. In T. Baum, M. Deery, C. Hanlon, L. Lockstone, & K. Smith (Eds.), *People and Work in Events and Conventions - A Research Perspective* (pp. 127-137). London: CABI.

DeFillippi, R. (2015). Managing project-based organization in creative industries. In C. Jones, M. Lorenzen, & J. Sapsed (Eds.) *The Oxford Handbook of Creative Industries* (pp. 268-283). Oxford: Oxford University Press.

DeFillippi, R., & Arthur, M. (1998). Paradox in project-based enterprise: The case of film making. *California Management Review, 40*(2), 125-139.

Drummond, S., & Anderson, H. (2004). Service quality and managing your people. In I. Yeoman, M. Robertson, J. Ali-Knight, S. Drummond, & U. McMahon-Beattie (Eds.), *Festival and Events Management - An international arts and culture perspective* (pp. 80-96). London: Routledge.

Duffield, S., & Whitty, S. J. (2015). Developing a systemic lessons learned knowledge model for organisational learning through projects. *International Journal of Project management, 33*(2), 311-324.

Earl, M. J. (2001). Knowledge Management Strategies: Toward a Taxonomy. *Journ al of Management Information Systems, 18*(1), 215-233.

Emery, P., & Radu, A. (2008). Conceptual paper: An exploration of time and its management for sport event managers. In M. Robertson & E. Frew (Eds.), *Events and Festivals – Current Trends and Issues* (pp. 103-121). London: Routledge.

Getz, D. (2002). Why festivals fail. *Event Management, 7*(4), 209–219.

Getz, D. (2018). *Event Evaluation*. Oxford: Goodfellow Publishers.

Getz, D. (2019). *Event Impact Assessment*. Oxford: Goodfellow Publishers.

Getz, D., & Frisby, W. (1988). Evaluating management effectiveness in community-run festivals. *Journal of Travel Research, 27*(1), 22-27.

Ghobadi, S., & Mathiassen, L. (2016). Perceived barriers to effective knowledge sharing in agile software teams. *Information Systems Journal, 26*(2), 95-125.

Hanlon, C., & Cuskelly, G. (2002). Pulsating major sport event organizations: a framework for inducting managerial personnel. *Event Management, 7*(4), 231-243.

Heisig, P. (2009). Harmonisation of knowledge management – comparing 160 KM frameworks around the globe. *Journal of Knowledge Management, 13*(4), 4-31.

Hinds, P. J., & Pfeffer, J. (2003). Why organizations don't 'know what they know': Cognitive and motivational factors affecting the transfer of expertise. In M. Ackerman, V. Pipek, & V. Wolf (Eds.), *Sharing Expertise: Beyond Knowledge Management* (pp. 3-26). Cambridge, MA: MIT Press.

Hislop, D., Bosua, R., & Helms, R. (2018). *Knowledge Management in Organizations - A Critical Introduction* (4th ed.). Oxford: Oxford University Press.

Husted, K., & Michailova, S. (2002). Diagnosing and fighting knowledge-sharing hostility. *Organizational Dynamics*, 31(1), 60-73.

Johnson, D. W., & Johnson, F. P. (2003). *Joining Together – Group Theory and Group Skills* (8th ed.). Boston A and B.

Junek, O., Lockstone, L., & Mair, J. (2009). Two Perspectives on event management employment: student and employer insights into the skills required to get the job done! *Journal of Hospitality and Tourism Management*, 16(1), 120-129.

Kodama, M. (2007). *Project-Based Organization in the Knowledge-Based Society*. London: Imperial College Press.

Lewis, J. P. (1998). *Team-Based Project Management*. New York: AMACOM - American Management Association.

Liebowitz, J. (2005). Conceptualizing and implementing knowledge management. In P. E. D. Love, P. S. W. Fong, & Z. Irani (Eds.), *Management of Knowledge in Project Environments* (pp. 1-18). Amsterdam: Butterworth-Heinemann.

Lindkvist, L. (2005). Knowledge communities and knowledge collectivities: a typology of knowledge work in groups. *Journal of Management Studies*, 42(6), 1189-1210.

Lockstone-Binney, L., Hanlon, C., & Jago, L. (2020). Staffing for successful events: Having the right skills in the right place at the right time. In S. J. Page & J. Connell (Eds.), *The Routledge Handbook of Events* (2nd ed., pp. 427-441). London: Routledge.

Maaninen-Olsson, E., Wismen, M., & Carlsson, S. A. (2008). Permanent and temporary work practices: knowledge integration and the meaning of boundary activities. *Knowledge Management Research & Practice*, 6(4), 260-271.

Mair, J. (2009). The events industry: the employment context. In T. Baum, M. Deery, C. Hanlon, L. Lockstone, & K. Smith (Eds.), *People and Work in Events and Conventions - A Research Perspective* (pp. 3-16). London: CABI.

Muskat, B., & Deery, M. (2017). Knowledge transfer and organizational memory: an events perspective. *Event Management*, 21(4), 431-447.

Pereira, L., & Gonçalves, A. F. (2017). *Knowledge management in projects*. Paper presented at the 2017 International Conference on Engineering, Technology and Innovation (ICE/ITMC).

Ragsdell, G., & Jepson, A. S. (2014). Knowledge sharing: insights from Campaign for Real Ale (CAMRA) festival volunteers. *International Journal of Event and Festival Management*, 5(3), 279-296.

Ragsdell, G., Espinet, E. O., & Norris, M. (2013). Knowledge management in the voluntary sector: a focus on sharing project know-how and expertise. *Knowledge Management Research & Practice*, 12(4), 351-361.

Rämö, H. (2002). Doing things right and doing the right things - Time and timing in projects. *International Journal of Project Management*, 20(7), 569-574.

Scarbrough, H., Swan, J., Laurent, S., Bresnen, M., Edelman, L., & Newell, S. (2004). Project-based learning and the role of learning boundaries. *Organization Studies*, 25(9), 1579–1600.

Smith, K., & Lockstone, L. (2009). Involving and keeping event volunteers: Management insights from cultural festivals. In T. Baum, M. Deery, C. Hanlon, L. Lockstone, & K. Smith (Eds.), *People and Work in Events and Conventions - A Research Perspective* (pp. 154-170). London: CABI.

Smith, K., Lockstone-Binney, L., & Holmes, K. (2019). Revisiting and advancing the research agenda for event volunteering. In J. Armbrecht, E. Lundberg, & T. Andersson (Eds.), *A Research Agenda for Event Management* (pp. 126-153). Cheltenham: Edward Elgar Publishing.

Stadler, R., & Fullagar, S. (2016). Appreciating formal and informal knowledge transfer practices within creative festival organizations. *Journal of Knowledge Management*, 20(1), 146-161.

Swan, J., Scarbrough, H., & Newell, S. (2010). Why don't (or do) organizations learn from projects? *Management Learning OnlineFirst*, 41(3), 325-344.

Toffler, A. (1990). *Future Shock*. New York: Bantam Books.

Van der Wagen, L. (2007). *Human Resource Management for Events – Managing the Event Workforce*. Amsterdam: Elsevier.

Van der Wagen, L., & White, L. (2014). *Human Resource Management for the Event Industry* (2nd ed.). London: Routledge.

Winkler, K., & Mandl, H. (2015). Knowledge management for projects. In M. Wastian, M. Wastian, M. A. West, & I. Braumandl (Eds.), Applied *Psychology for Project Managers* (pp. 73-84). Springer.

3 Knowledge Management Activities, Models and Frameworks

Learning objectives

- Define and discuss key knowledge management activities.
- Understand the role as well as the limitations of ICT for each of the knowledge management dimensions.
- Understand the particular importance of knowledge sharing in event organisations.
- Explore different knowledge management models.
- Apply knowledge management frameworks to event organisations.

Introduction

The knowledge-based theory of the firm is centred on two key principles: first, knowledge can be a source of competitive advantage, especially if it is difficult to replicate and copy, if it is organisation-specific, and directly related to the organisation's products, services or processes. Second, the sharing of information and knowledge within organisations is believed to be more effective than within markets (Hislop et al., 2018). Furthermore, if knowledge is considered the most important asset for achieving competitive advantage in an organisation (where competencies, capabilities and skills are more important than other types of assets and resources), then the organisation can be defined as 'knowledge-intensive'. Or, as Nonaka and Takeuchi (1995: 7) famously argued, "the future belongs to people who use their heads instead of their hands." Knowledge-intensive firms can be described as companies with work that is intellectual and where the workforce consists of well-qualified employees. They can also be different to other, non-knowledge-intensive firms, in terms of their structure (i.e., less hierarchical), the character of their workforce, products and services (e.g., non-standardised products, high quality services, such as memorable experiences), as well as the nature of the work processes themselves (i.e., complex, creative, requiring collaboration between different teams and departments) (Alvesson & Kaerreman, 2001; Hislop et al., 2018). Taking these characteristics into account, event organisations can to some extent be classified as 'knowledge-intensive' and it is therefore important to effectively manage knowledge management activities and processes within these organisations.

Hislop et al. (2018: 79) nicely summarise a number of sources from the wider knowledge management literature that define an organisations' intellectual capital to consist of three elements, which together constitute the sum of all the knowledge used to gain organisational competitive advantage:

- *Human capital:* consists of all the knowledge, skills, abilities, and expertise residing in and used by individuals within the organisation;

- *Organisational capital*: is the collective institutional knowledge and organisational memory which has been codified/made explicit in databases and other repositories, systems, processes, and structures; and

- *Social capital*: is all the knowledge embedded in and available through personal relationships (networks between people) within the organisation. This implies that a certain element of trust between people needs to exist in order for them to share knowledge with each other. In other words, social capital is more than the sum of all the individual employees' capital, it is according to Dalkir (2017), the glue that holds them together.

All three types of capital underpin the process of knowledge management and are equally important. This chapter will first cover the most commonly identified activities in knowledge management: the creation of new knowledge; the identification and acquisition of knowledge for individuals, teams and the organisation as a whole; the implementation and utilisation of knowledge in day-to-day practices; the sharing and transfer of knowledge with others (inside and outside the organisation); and finally, the storage of knowledge in order for it to be documented and hence, made explicit. The role of ICT systems as well as their limitations will be discussed throughout each of these dimensions of knowledge management. Where applicable, examples from the events and festival literature will be presented at the end of each section to show how some of these knowledge activities have been explored and investigated in an events context. The second part of the chapter summarises some of the best-known knowledge management frameworks built around these key knowledge management activities: it starts with Nonaka and Takeuchi's (1995) knowledge creation spiral model, then discusses Wiig's (1995) knowledge management model and finally McElroy's (2003) knowledge life cycle framework. Each framework or model will be applied to event organisations and event-related examples throughout.

Key knowledge management activities

According to Heisig (2009) most knowledge management frameworks in the 1990s and early 2000s identified six key knowledge management activities: create, identify, acquire, use, share and store knowledge (with some using synonyms). Each of them will be discussed more in-depth below, and the supporting role of ICT in each knowledge activity will also be briefly summarised. It is commonly acknowledged that technology can never replace knowledge; it is merely a channel or medium for conveying, sharing and storing data and information (Halawi et al., 2017). Furthermore, it is interesting to note that out of the six knowledge management activities, knowledge sharing has by far received the most attention in both academic literature and management practice so far. Over the last 10 years, these trends have continued, but are now embedded in a more practice-based understanding of knowledge management (see Chapter 4). Table 3.1 below summarises typical knowledge management activities and some of the synonyms used in the literature as well as similar activities, which can be classified under the same category:

Table 3.1: Knowledge management activities

KM activity	Synonyms and similar activities
Create	Generate, develop, innovate, build/sustain, produce, evolve
Identify	Organise and classify, structure, analyse, determine, review, locate, investigate, discover, survey and categorise, map, find
Acquire	Collect, import, provide, get, source, gather
Use	Apply, leverage, re-use, enable, exploit, capitalise, deploy
Share	Transfer, distribute, communicate, collaborate, diffuse, disseminate, allocate, network, cooperate
Store	Retain, capture, codify, package, secure, archive, document, maintain, preserve, protect, accumulate

Knowledge creation

It is important to note that event organisations produce customised, specifically designed services, rather than off-the-shelf products and services. This means that new knowledge must constantly be created. Knowledge creation is thereby seen as the key process for innovation and hence building competitive advantage (Nonaka & Takeuchi, 1995). Creating knowledge includes activities such as, producing and developing new ideas, generating new knowledge for specific tasks, as well as organisational evolving and innovating. In order for knowledge to effectively be created in an organisation, according to Nonaka and Takeuchi (1995: 73-83), several conditions can enhance the process:

1 **Intention**: Intention is an organisation's aspiration to its goals. It can be expressed in an organisation's vision, mission and values and provides the most important criterion for judging the relevance and truthfulness of a given piece of knowledge. Any newly created knowledge will also be based on the organisation's aspirations and should effectively contribute to it.

2 **Autonomy**: Knowledge is most effectively created in organisations where individual employees have autonomy to act and make decisions. This level of autonomy and the extent of it is, of course, situated and framed within the first factor, the organisational intention.

3 **Fluctuation and creative chaos**: These two conditions stimulate the way the organisation acts with regards to its external environment. Any fluctuation in a business environment generates the conditions

for creative chaos to happen, which in turn starts the process of knowledge creation in an attempt to respond to the external environment factors. For example, creative chaos usually occurs when the organisation is facing some sort of crisis (e.g. a decline in performance due to rival organisations rapidly gaining competitive advantage). Tension within the organisation then increases, which forces employees to focus on creating new knowledge to solve the problem. In other words, individuals and organisations tend to be more creative when some external stimulus forces them to have to rethink the way they view the world, their taken for granted mental models, and knowledge processes;

4 **Redundancy**: Redundancy in terms of knowledge creation can be a very positive factor. It occurs when information goes beyond the immediate operational requirements of organisational members or where it intentionally overlaps and spans a range of business activities, management responsibilities, and the company as a whole. It can be beneficial when an organisation is trying to break down silos and groupthink, and can, for example, be achieved through job rotation. Employees rotating from one job to another will start to make connections between different departments, hierarchical levels, and job tasks. They can create new knowledge from using the same kind of knowledge in different contexts and will notice how some knowledge needs to be shared with everyone, whereas other types of knowledge may only be relevant to a particular team or function; and lastly,

5 **Requisite variety**: The final condition for the effective creation of knowledge in an organisation is requisite variety. This means that an organisation's internal diversity must match the variety and complexity of its external environment. It will allow employees to have all the information available for creating new knowledge that is in line with the demands of its external environment.

Knowledge creation will be a specific model further discussed later on in this chapter (see page 54), but it is important to acknowledge how ICT can enhance the process of knowledge creation in an organisation in the sense that it can mediate communication (Halawi et al., 2017), for example, when employees contribute to a forum where ideas, problems, and interpretations are shared. These systems can initiate dialogue and interaction among employees (both online and offline) and help them interpret information and create new insights together, rather than simply on their own (Alavi & Leidner, 2001). This, however, depends on individuals contributing to the system on a regular basis. Some employees might feel reluctant to do so or might not find the time to contribute – a particular issue in the fast-paced event environment.

Knowledge identification

The second knowledge management activity, knowledge identification, is the process of proactively identifying internal organisational knowledge, or in other words, the knowledge management process whereby organisations take steps to identify the relevant and needed knowledge that exists within their boundaries (Tow et al., 2015). It is sometimes also referred to as Knowledge Audit or Knowledge Sourcing and includes activities such as locating, finding and discovering knowledge; mapping, organising and classifying knowledge; as well as structuring, analysing and reviewing existing knowledge within an organisation. Within the wider knowledge management literature, the process of knowledge identification has so far received the least attention out of all knowledge activities. This is partly because of knowledge identification being closely related to other knowledge activities and sometimes difficult to separate from them. For example, in order for knowledge to be shared, it first needs to be identified. Knowledge identification and knowledge sharing therefore always go hand-in-hand.

There are several methods and tools for identifying knowledge, such as Knowledge Sharing Systems, where employees write down what they know, or Knowledge Mapping, which is a graphical representation of who knows what within the organisation and where that knowledge is stored. For an example of how Knowledge Mapping can be used in an event organisation, see Singh et al. (2007). Again, the overlap with other knowledge activities (knowledge sharing, knowledge storage) is evident here. ICT can further enable knowledge identification, for example, through the use of intranets where mutually shared organisational knowledge can be stored, categorised and structured, as well as reviewed (Turulja & Bajgoric, 2018). This can be particularly useful if it is integrated both horizontally and vertically (Alavi & Leidner, 2001). However, as previously noted, this process is time consuming.

Furthermore, only certain types of (explicit) knowledge can be stored and hence identified on the system, while the actual acquisition or use of the knowledge later on might be more complex (tacit) and therefore difficult to document on the system.

Knowledge acquisition

Knowledge acquisition from individuals or groups can been defined as, "the transfer and transformation of valuable expertise from a knowledge source (e.g., human expert, documents) to a knowledge repository (e.g., corporate memory, intranet)" (Dalkir, 2017: 123). It includes activities such as sourcing knowledge that is relevant to a certain problem faced, as well as collecting and gathering knowledge (both explicit and tacit) and importing it into other business processes. Knowledge acquisition can be achieved through a number of different approaches, some might be very formal approaches in certain organisations, while in other organisations they can be more informal. Three of the most common approaches to knowledge acquisition are: interviewing experts; learning by being told; and learning by observation. In the case of capturing tacit knowledge, usually a combination of all three approaches is required. Furthermore, all three approaches to knowledge acquisition can occur at the individual level, but also at the group and, in some cases, even the organisational level, where the acquired knowledge can then be added to the organisation's memory or repositories (Dalkir, 2017).

Interviews are the most common method for acquiring knowledge from an expert. Through conversation and asking questions, the expert can reveal key information, processes and practices they think about, can explain how these are related to other business processes, as well as clarify the steps they go through when making a decision or solving a problem. In an event organisation, for example, a new staff member can interview a senior staff member about the process of applying for a Government grant for funding. The senior staff member, who has done this many times and is an expert on writing funding applications, will be able to explain how grant applications relate to other areas of event management (budgeting and finance, marketing, PR), the specific steps they go through when writing the application (key information that needs to be provided in a clear and succinct way), as well as the process they go through when trying to solve a problem that might arise (data relevant to the application might be missing and needs to be analysed further before being able to submit the application). The new staff member can acquire some of the tacit knowledge necessary for writing a grant application through asking the right questions and probing further when something is unclear.

The second common approach to knowledge acquisition, learning by being told, often happens in conjunction with interviewing an expert and can help to refine the acquired knowledge and directly apply it to a work task. In this case, the new staff member in the example above would listen to the expert talking them through the process of writing a grant application, but it would be down to the new staff member to actually go through the process, with the expert telling them what to do during each step along the way. The third approach, learning by observation, is again closely related to the other two approaches and is about acquiring tacit knowledge through observing the expert solving a problem. For example, the new staff member who is working through the grant application, could present a 'what if' scenario to the expert and ask them to solve the problem so that they can watch and learn the best way to approach this type of scenario.

Knowledge can of course also be acquired through official induction sessions and training programs. Training should aim to provide information and knowledge to new employees so that they learn to perform their job effectively (Wiig, 2004). This includes formal training sessions, as well as on-the-job training with other staff members, experts, or in teams. Induction manuals, how-to guides and other forms of explicit/documented knowledge can help here, but it is crucially important to also use induction and training sessions to teach employees who is who, who knows what and how things are done within the organisation. Employees can therefore acquire knowledge about what is acceptable and not acceptable within the organisation; its vision, mission and value systems; as well as other 'soft' factors such as how to communicate with others, how to best work in a team, and how to contribute to the organisation and its aims more broadly. ICT is mainly used here to support these activities, for example through providing electronic manuals and how-to guides that can easily be accessed and also be kept up-to-date.

Knowledge use

Needless to say, any knowledge that has been created and identified within the organisation and acquired by individuals or teams, should ideally then be regularly used or applied to work tasks, otherwise it will be lost again. Knowledge use therefore includes activities such as, applying knowledge, exploiting and capitalising on the knowledge created and identified, as well as re-using knowledge in different situations and at different times. Employees apply knowledge to their day-to-day tasks, as well as use knowledge when making decisions and solving problems. They can therefore, according to the 'all work is knowledge work' perspective (see Hislop et al., 2018), be called 'knowledge workers.' Hislop et al. (2018: 75, emphasis added) defined knowledge workers as, "[a]nyone whose work involves the _use_ of a reasonable amount of tacit and contextual and/or abstract/conceptual knowledge."

At an individual level, the personal characteristics of the employee who is aiming to use or apply some form of knowledge tend to play an important role in terms of how effectively they are going to achieve this. The knowledge worker's personality type, their preferences of how they want to learn, as well as how they want to receive information, will have an impact on how effective they are in finding, understanding and then using the knowledge (Dalkir, 2017). This includes their cultural background, whether they are a visual learner or prefer learning-by-doing, and their mental capacity when putting knowledge to work. It is therefore important to note that while official training programs can help some employees learn and understand how to use certain knowledge, it might not be the most effective way for everyone in the organisation. Others might prefer a trial-and-error approach, where they use and re-use knowledge in different ways and through this process learn which way is the best, most efficient, and most effective over time. Furthermore, the type of knowledge being required might also determine the way it is used. For example, when explicit knowledge is used to solve a problem, a simple email to confirm the task might be enough. Whereas if tacit knowledge needs to be used, the best way of using it will most certainly depend on the context and situation – what works today might not work again tomorrow.

Lastly, knowledge use also includes activities of re-using knowledge, which are particularly important to acknowledge in an events context.

Re-using knowledge requires the ability to recall, and in many cases reflect on past experiences and past use of the knowledge in a similar situation (Dalkir, 2017). This can be difficult to achieve in an event organisation where staff members join and leave the organisation at different times, and where a lot of knowledge only needs to be used once a year or even less often. Having a small team of permanent staff members available in the organisation, who recall how the knowledge was created in the first place, how it was identified and acquired, can help new (seasonal) staff members understand how to best use this knowledge in a certain situation or context. Within the events and festival literature, Blackman et al. (2017) recently investigated processes of capturing/acquiring and using/reusing knowledge at the 2010 Vancouver Olympic and Paralympic Games. They argued that these activities can greatly contribute to the development of human capital in mega-sport events as well as for the host community, but found that this was not successfully done at the Vancouver Olympics and a lot of knowledge was therefore lost rather than used and reused after the event.

It has been argued that in some cases, ICT can enhance the speed of knowledge use and application, especially when the knowledge can be codified and applied to organisational routines. Workflow automation systems are a good example here (see Alavi & Leidner, 2001), however, it must be stressed that in event organisations, there is hardly any routine work, and this is therefore difficult to implement.

Knowledge sharing and knowledge transfer

As previously mentioned, the activities of knowledge sharing and knowledge transfer have by far received the most attention in the academic literature. The same is true for research on knowledge management in events: most studies focus on knowledge sharing activities and processes, particularly with new seasonal staff members and volunteers (Stadler, 2019). Knowledge sharing includes activities such as, communicating knowledge with others inside and outside the organisation; disseminating, diffusing and distributing knowledge to those who need it; as well as networking, collaborating and cooperating with others. People are at the heart of knowledge sharing activities, as Dalkir (2017: 168) argued:

It turns out that, not only are other people the preferred source of information, but that there are a number of reasons for this. One is, of course, that it is often faster, but this is not the only reason. When we turn to another person, we not only end up with the information we were looking for, but we also help learn where it was to be found. The other person may help us to reformulate our question or query, tell us where we were on the right track and where we strayed, and, last but not least, that the information is coming to us from a known and usually trusted, credible source. In other words, people are the best means of getting not only a direct answer but also 'metaknowledge' about our search target and our search capabilities. Talking to other people provides a highly valuable learning activity that is primarily a tacit-tacit knowledge transfer, as this type of knowledge is seldom rendered explicit or captured in any form of document.

In its most basic sense, knowledge sharing can be explained using the following steps: knowledge sharing starts with an independent sender who holds some sort of knowledge; this knowledge is then transferred from the independent sender – via a transmission channel or mechanism – to a separate receiver. The process occurs within a certain context, and the success of the process depends on three factors: 1) the sender is knowledgeable and also willing to share their knowledge; 2) an appropriate channel or mechanism for the transfer is used, depending on the type of knowledge being transferred (e.g. codified knowledge can be transferred via email, tacit knowledge can be transferred in meetings or through learning-by-doing); and 3) the receiver has the capacity to absorb, learn and use the new knowledge (Hislop et al., 2018).

Sharing knowledge can be intrinsically motivating for individual members of the organisation and it can also be beneficial at a group level, such as for enhanced team performance. Other motivating factors include status enhancement, for example, becoming the known expert on a specific task. In some cases, there might even be extrinsically motivating factors, such as a reward, bonus or promotion (Hislop et al., 2018). However, at the same time, sharing knowledge can be very time consuming and some employees might feel that they are giving away some of their individual status and power, leading to them hiding or hoarding their knowledge rather than sharing it – an issue already discussed in Chapter 2.

People, however, are not the only means of transferring knowledge. Wiig (1995: 344-345) further summarised key knowledge sharing activities in organisations as follows:

- *Person-to-person knowledge transfers:* as discussed, this is the most frequent approach to transferring knowledge, and can be achieved through, for example apprenticing; working in teams; consulting with experts; and formal and informal discussions. It is important to note that many of these activities cannot be 'managed', as they quite often 'just happen.'

- *Passive communications and repositories, or in other words, 'documents':* these more traditional approaches include, for example, knowledge sharing through textbooks, reference manuals, or company reports.

- *Education and training programs*: these can be, for example, personal, informal one-to-one interactions, classroom programs based on well-structured materials, videotapes and recorded lectures and seminars. Creating these programs in the first place might be quite expensive and time-consuming, but can later help in sharing key knowledge with a large number of people in a succinct and easy-to-understand format.

In terms of the use of ICT to support knowledge sharing and transfer, it has mainly been applied to informal means (such as Lotus Notus discussion databases) and formal means (such as corporate directories) (Turulja & Bajgoric, 2018). A good example provided by Alavi and Leidner (2001) is the use of a directory of employee interest profiles, where expertise and skills, and hence experts, can be identified. These can then be approached directly or added to email lists, share points, shared drives and clouds. ICT in this sense can therefore be used as a communication and collaboration tool (Halawi et al., 2017). It can be particularly useful when employees tend to only work in their usual teams, develop groupthink, and can therefore benefit from contacting somebody outside their immediate line of communication. This could help them develop a different perspective or challenge their thinking. Electronic billboards and discussion groups can also initiate and facilitate contact between somebody in need of a certain piece of knowledge and an expert possessing this knowledge (Alavi & Leidner, 2001). Knowledge can then be shared quickly and more effectively.

Not surprisingly, knowledge sharing has received the most attention out of all knowledge activities within the events literature so far. The following examples summarise some of these findings: Within a volunteer-led festival organisation investigated by Ragsdell and Jepson (2014), learning-by-doing approaches were the most successful ones in terms of sharing and transferring knowledge within the organisation. In terms of motivational factors for knowledge sharing within this organisation, Ragsdell and Jepson (2014) highlighted trust in the management of the event and in the quality of project knowledge, as well as pride in doing a good job as absolutely key. Another study by Clayton (2016) also looked into knowledge transfer amongst volunteers at UK festivals. She found that knowledge sharing and transfer activities were particularly successful when volunteers were motivated and engaged, which then leads to repeat volunteering and hence successful reuse of knowledge rather than starting from scratch with a new team of volunteers each time and having to reinvent the wheel.

Stadler et al. (2014) looked into KM roles and responsibilities (these will be further discussed in Chapter 5) within an Australian music festival and advocated an inter-disciplinary team structure rather than functional teams working in silos. These inter-disciplinary teams enhance the process of sharing knowledge both within each individual team, as well as between different teams and therefore the organisation as a whole. Lastly, Muskat and Deery (2017) explored knowledge transfer activities between paid staff members and volunteers during each of the three phases of pre-, during and post-event. Within small- and medium-sized event organisations, they found differences in the type of knowledge being shared at each of the stages: at the pre-event stage, it was mainly explicit knowledge that was shared; during the event, new tacit knowledge was created and to some extent shared, but not always successfully converted into explicit knowledge during the post-event stage – an issue discussed further in the following section, which leads to the organisation not being able to build an organisational memory over time.

Knowledge storage

The last of the six knowledge management activities, knowledge storage, is important for any organisation in order to create an organisational memory. This includes information and knowledge in the form of, for example, written documents, electronic databases, documented organisational processes and practices, as well as tacit knowledge held by individuals and networks of individuals (Alavi & Leidner, 2001), and in the case of event organsiations, forms part of the event evaluation process (Getz, 2018). It can usually be stored in knowledge repositories, which contain valuable content (both explicit and tacit knowledge), experiences, shared understanding and know-how from people who are, or indeed used to be, part of the organisation. This knowledge has been tried and tested many times, and has been found to work for that particular organisation (Dalkir, 2017). In other words, knowledge storage includes activities such as archiving and preserving knowledge; codifying knowledge in order to make it explicit; as well as packaging, securing and protecting it to avoid knowledge leakage.

ICT systems (particularly knowledge storage in data warehouses, multimedia databases, or document repositories) can be of use here, especially when used in conjunction with sophisticated retrieval techniques, such as query languages of database management systems. They can significantly increase the speed of accessing organisational memory (Alavi & Leidner, 2001; Turulja & Bajgoric, 2018), but have so far only been used in large firms and multinational companies. They are too expensive and resource-intensive for SMEs such as event organisations. Furthermore, as previously discussed, time is limited once the event is over and not everything learned can be made explicit and stored at this stage.

The role of technology in information and knowledge storage has been identified as very important for mega-events and a few studies have specifically looked into this: Halbwirth and Toohey (2001) and Toohey and Halbwirth (2005) investigated the Sydney 2000 Games Information System as a specific knowledge project and found some success in its use. However, the key element missing in the ICT system is an acknowledgement of the organisation's environment and culture. In order for such a system to be effective in supporting knowledge storage activities, it needs to be embedded in the organisation's culture and value system. Singh and Hu (2008) on the other hand, explored knowledge documentation and transfer processes between the Athens 2004 Organising Committee and the Greek National Tourism Organisation. It became evident in their study that both institutions documented and stored some of the vast amount of knowledge created and were even able to share some of it with each other. However, once again, these activities were limited to the storage and transfer of only explicit types of knowledge.

Knowledge management frameworks

Based on the knowledge management activities discussed above, in the 1990s and early 2000s there was a rapid increase in the development of knowledge management models, frameworks and theories both in the academic literature, as well as applied in practice (see Heisig, 2009, 2014). Many conventional frameworks and models from these early days of knowledge management (first generation, as discussed in Chapter 1) were based on a few key assumptions, such as (see McElroy, 2003: 5):

- "It's all about getting the right information to the right people at the right time.

- If we only knew what we know.

- We need to capture and codify our tacit and explicit knowledge before it walks out the door."

According to Hislop et al. (2018), for example, the basic model for putting any kind of knowledge management system in place consists of three key steps:

1 Create a knowledge management initiative and identify what kind of knowledge is important for which process, then make it explicit;

2 Collect all the explicit (codified) knowledge and save it in a central repository, where the knowledge is structured in a systematic way; and

3 Make the knowledge accessible to those people in the organisation that need it.

It is beyond the scope of this book to summarise all knowledge management frameworks, but the three models discussed in this chapter are among the most cited in the literature: Nonaka and Takeuchi's (1995) knowledge creation spiral, Wiig's (1995) four knowledge flow functions, and McElroy's (2003) knowledge life cycle framework. They will be applied to event-related examples throughout to demonstrate how they can be used in practice.

Nonaka and Takeuchi's knowledge creation spiral

In an attempt to describe the difference between Japanese and Western firms in dealing with knowledge management, Nonaka and Takeuchi (1995) came up with a complex model of knowledge creation. Even though it is based on the management of large firms, it is a very useful model for any organisation because it includes both explicit and tacit knowledge, the conversion processes from one to the other, as well as the 'soft' factors of people and organisational culture. The model provides an overview of the entire knowledge management process. The authors argue that knowledge creation is essential for continuous innovation, and thus competitive advantage in the long run. Their model is regarded as a classic in the knowledge management literature. The core processes are still used in their original form, but over the years the model has been amended, added to and improved (see Nonaka & Konno, 1998; Nonaka et al., 2000; Nonaka & Toyama, 2003).

Nonaka and Takeuchi (1995: 3) define knowledge creation as "the capability of a company as a whole to create new knowledge, disseminate it throughout the organization, and embody it in products, services and systems." It therefore includes the process of creating knowledge as well as sharing knowledge within the organisation, and both explicit and tacit knowledge. The authors describe four modes of knowledge conversion: socialisation, externalisation, combination and internalisation.

1. *Socialisation* deals with the process of connecting tacit knowledge with tacit knowledge, which includes the sharing of experiences and creating shared mental models and technical skills. Tacit knowledge is thereby acquired not through the use of language, but rather through observation, imitation, practice, and hence experience, such as for example through on-the-job training. In an event organisation this could be achieved through, for example, a new staff member observing long-term staff members when interacting with sponsors. Passing on information on *how* to best approach a new sponsor is not enough, it needs to be experienced and learned in a certain context.

2. *Externalisation* means making tacit knowledge explicit, that is tacit knowledge "taking the shapes of metaphors, analogies, con-

cepts, hypotheses, or models" (Nonaka & Takeuchi, 1995: 64). This process is enhanced through dialogue and collective reflection. Metaphors in an event organisation are sometimes created in, for example, the staff room. The technical team, for instance, can come together here, discuss and reflect, and together create metaphors for different work practices that carry with them a certain meaning and help make tacit knowledge explicit. It is important to note that these metaphors, however, will only make sense to the people involved in the process and cannot easily be transferred into other departments or teams.

3   The third part of the knowledge creation process combines different types of explicit knowledge and is thus called *combination*. Here, documents, meetings, telephone conversations, and e-mails are essential for creating new explicit knowledge. Most event organisations use databases, which are also important tools for combining different types of explicit knowledge, such as basic attendee information (demographic information, contact details) with records of how and when they have been contacted (email lists, telephone conversations), and what kind of information has been shared with them (e.g., promotional email, online ticket booking, follow-up telephone conversation). This explicit knowledge grouped together provides more specific insights than the sum of its parts and is hence an important element of the knowledge creation process.

4   *Internalisation*, finally, covers the process of transforming explicit knowledge into tacit knowledge. Documents, manuals, or shared stories are used to help individuals create new tacit knowledge in the form of, for example, shared mental models or technical know-how. In other words, when employees use explicit organisational knowledge (e.g., operating manuals), they absorb it into their day-to-day work practices, where it shapes their tacit practices. An important aspect of this part of the internalisation process is learning by doing, but it can also be achieved through, for example, sharing success stories. Event organisations can benefit from telling and retelling past success stories during an induction session or training for new staff members. This helps make the experience come alive and therefore creates a shared tacit mental model of 'how things work' in the organisation.

It is important to note that these four modes of knowledge conversion do not start or end at a certain point; rather, they form a dynamic process of knowledge creation. All four elements are equally important. Nonaka and Takeuchi (1995) describe the process as a knowledge creating spiral, which helps the organisation to continuously improve and learn. Figure 3.1 describes the entire process, and also shows how the three levels of individual (i), group (g), and organisational (o) knowledge are related to the modes of knowledge conversion: socialisation tends to happen between individual employees, whereas externalisation occurs when individual team members start sharing their tacit knowledge with the group. Combination is an important process involving several groups, and lastly, internalisation brings back organisational knowledge to an individual level.

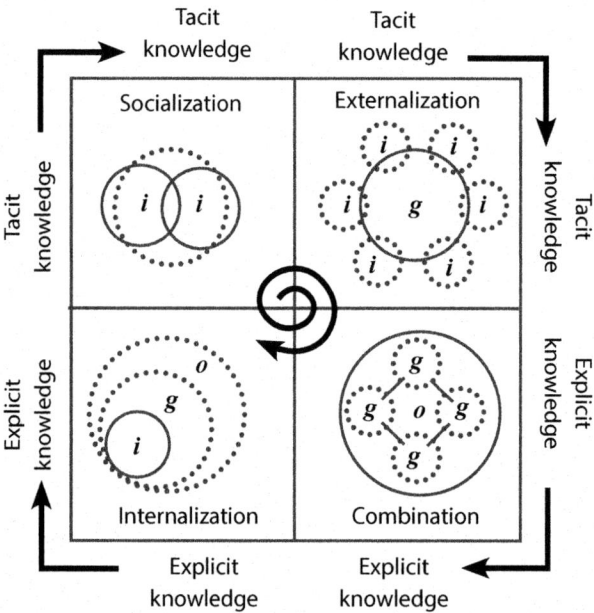

Figure 3.1: Spiral evolution of knowledge conversion. (Nonaka & Konno, 1998: 43), republished with permission of California Management Review, permission conveyed through Copyright Clearance Center Inc.

Table 3.2 (adapted from Nonaka & Takeuchi, 1995, and Hislop et al., 2018) summarises the different types of knowledge conversion, the level of where this conversion takes place, and provides an event-related example for each.

Table 3.2: Four modes of knowledge conversion in events

	Socialisation	Externalisation	Combination	Internalisation
Type of knowledge conversion	Tacit → tacit	Tacit → explicit	Explicit → explicit	Explicit → tacit
Change in level of knowledge	Individual – individual	Individual – group	Group – organisation	Organisation – individual
Event-related example	Observing an experienced staff member interacting with a new sponsor	Creating metaphors in the staff room or behind the stage	Database recording attendee contact information and communication	Telling, retelling and reliving stories of past event highlights and successes

Nonaka and Takeuchi's (1995) model is arguably the most influential framework in the wider knowledge management literature. It has been applied by Blackman et al. (2017) to knowledge transfer activities between volunteers at the 2010 Vancouver Olympic and Paralympic Games in combination with the Knowledge Value Chain framework. They argue that through the use of these knowledge management frameworks, human capital legacy can be developed in the long-term, which gives mega-event organisations a competitive advantage. The study shows that while human capital legacy was an anticipated outcome identified by the organisation, there was no specific strategic plan in place for how this can be achieved in an effective and efficient way. The use of Nonaka and Takeuchi's (1995) model in combination with the Knowledge Value Chain framework was therefore recommended in order to make these processes more explicit.

It is important to note, however, that the socialisation-externalisation-combination-internalisation model has also been critiqued by many (see Hislop et al., 2018: 110), particularly in terms of its unconvincing empirical evidence to support the theory. The examples presented in Nonaka and Takeuchi's (1995) book have been regarded as brief and anecdotal, and hence unconvincing. A further critique comes from the distinction between explicit and tacit knowledge, where other theorists argue that there is always an element of tacit knowledge to any form of explicit knowledge, and also that tacit knowledge can never be fully made explicit and the process of externalisation is therefore limited.

Lastly, the third main critique of the model is based on the fact that the theory was developed in a Japanese context and is hence not universally applicable to other organisations and companies using different (Western) business practices. The values, work practices and cultures are arguably very different in Japanese companies and cannot be generalised to other parts of the world. Despite these criticisms, however, it remains the most referenced framework in the literature and some of the ideas are undoubtedly interesting to explore and apply further in an events context.

Wiig's four knowledge flow functions

Wiig (1995) provides another classic knowledge management framework and argues that in order for a company or organisation to work at its best, we need to understand how knowledge flows within the organisation. He developed a Knowledge Flow Analysis model consisting of four main knowledge flow functions: build knowledge, organise and hold knowledge, distribute and pool knowledge, and apply knowledge. A Knowledge Flow Analysis thereby aims to "find and describe opportunities for improving knowledge flows around selected areas of the organization in order to do business better" (Wiig, 1995: 212). It is about identifying and investigating the paths, means and utility of all (aggregated) knowledge flows within an organisation, rather than how knowledge flows between individual employees. Wiig (1995) argued that even when work processes seem to be effective and efficient, knowledge does not always flow properly or timely, or even to the areas that most need it. Opportunities for learning and improvement can therefore easily be missed, or in some cases, employees might not use the best knowledge available when making decisions because it has been lost, forgotten or simply overlooked. Knowledge Flow Analysis therefore helps to perform work tasks better, improve work processes and functions, and hence also improve products and services. In other words, knowledge flows can be positive or negative; it is the aim of a Knowledge Flow Analysis to identify the positive ones and make them more explicit.

Figure 3.2 depicts a basic overview of the model applied to events and a description of each function is provided further below.

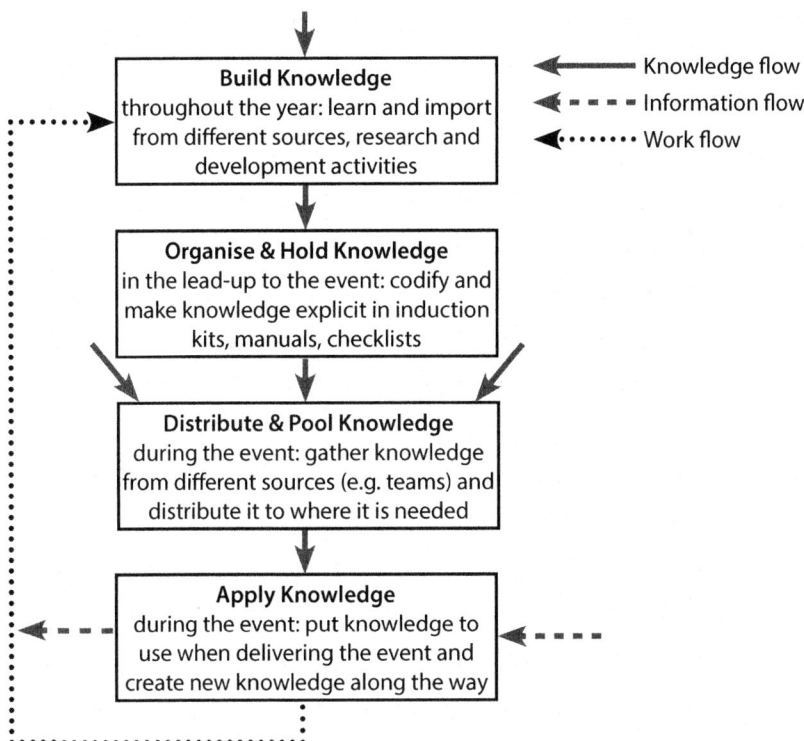

Figure 3.2: The four major knowledge flow functions applied to events. (adapted from Wiig, 1995)

The four functions can be further described as follows (see Wiig, 1995: 219-220):

1 *Building knowledge*: knowledge can be built through learning, research and development activities, innovation, or simply through importing knowledge from existing sources. In an event organisation this can be an ongoing process throughout the whole year, mainly performed by the permanent staff and directors.

2 *Organising and holding knowledge*: knowledge that has been codified and made explicit is being held in books, manuals, databases, and organisational memory. If it is further organised, then it can be made available for specific purposes. Event organisations can, for example, benefit from organising knowledge in the lead-up to the event, such as in manuals, induction kits or checklists, in order for it to be readily available when needed.

3. *Distributing and pooling knowledge*: here knowledge is pooled or assembled from different sources (e.g., different members of a team, or expert networks) then distributed to where it is needed or being used. This is perhaps the most important function within an event organisation, and probably prevalent during the actual event. If knowledge is pooled from different teams and distributed to where it is needed, decisions can be made quickly and efficiently.

4. *Applying knowledge to work objects*: knowledge is put to use when improving existing products and services or when creating new ones. Again, during the event itself, existing knowledge is constantly put to use, while at the same time new knowledge is also created, which can then feed back into the first function (post-event).

It is important to note, however, that these steps are not necessarily always sequential. Especially in a 'pulsating' event organisation, where people come and go at different points in time, these steps might overlap or be performed at different times for different teams. The following key questions were identified by Wiig (1995: 221) as important to constantly ask when going through the process (adapted here and applied to an events context):

- How does knowledge flow from the knowledge pool to the design of the event experience?
- Who contributes the knowledge that will be used? Are there others who may have better or complementary knowledge?
- What are the knowledge contributions of the different parties? e.g., are only experts in charge or do new staff members and volunteers also get a chance to contribute?
- Who are the individuals who learn from experience on-the-job? What are their specialties? Are they in a position to feed back what they learn to improve the quality of work and otherwise help improve event operations?
- What are the different routes and processes for learning and knowledge building and other kinds of knowledge transfer?
- How is knowledge fed back to the functional tasks? e.g., to improve the way it is applied to event marketing and PR; to improve the design of the event experience?

- How can knowledge be used to work smarter?
- What are the opportunities to use knowledge to improve unique event experiences for external and internal customers?
- Where and how is good existing knowledge stored so that it is not forgotten, and how can it be accessed?
- Who monitors and controls the quality of knowledge, weeding out outdated and wrong knowledge – especially once the event is over? e.g. permanent staff members, executive team?
- Is any knowledge being lost or getting into the hands of competitors? e.g. when seasonal staff members and volunteers move on after the event is over?

Finally, Wiig (1995) states that once the Knowledge Flow Analysis has been completed, emphasis can be put on working on four knowledge flow dimensions. These include, 1) do the job; 2) do the job better; 3) improve the work function and work practices; and 4) improve products and services. In the first step, 'Get the Job Done!' simply involves the application of relevant knowledge to produce high quality products or services. Second, 'Do the Job Better!' is all about bringing in new operational knowledge in order for employees to perform their tasks and processes better. This includes teaching employees the principles and operating objectives that underlie their job function, so that they are then able to handle exceptional situations and last-minute changes. Third, 'Improve the Work Function and Work Practices!' requires a broader view of the entire system and means using knowledge to change the system of production and/or service. Last 'Improve Products and Services!' is about building, organising, using, and applying knowledge to improve existing products and services or to create new ones through knowledge embedded in the above work functions and practices.

In order to explain this further in the context of event organisations, the following example will be used: a small-scale, local community festival is working on finalising the program and promoting it to a wider audience. The marketing team has been tasked with designing posters and fliers, as well as visuals and content for the website and various social media channels. The 'work object' for Wiig's (1995) framework that the marketing team is working on, can therefore be defined as the main 'festival poster design.' Referring back to the four dimensions discussed, this process and the flow of knowledge could therefore be described, and changes could be suggested as follows:

1. *Get the Job Done!* The marketing team are fully knowledgeable of the theme of this year's festival, as well as the main program elements and the main sponsors. They can design a poster along those lines using their expertise, skills, and pooled knowledge within the team. Important components within this knowledge flow dimension therefore include people, artefacts, work aids, and specific activities of the work function.

2. *Do the Job Better!* About two weeks before the posters need to go out for printing, a production meeting is being held with the two main headline acts of the festival, which creates a difficult situation: The creative team decide to slightly change the theme of the festival (in line with what the headline acts had in mind, emphasising the local history and traditions) in order to make it more relevant and accessible to a local audience. This new knowledge is relayed back to the marketing team, as they will need to redesign the poster accordingly. As part of the 'do the job better' process within the organisation, the marketing team have learned to network with peers, suppliers, customers, and other stakeholders with whom they now need to collaborate to perform the assigned tasks and redesign the poster effectively and efficiently.

3. *Improve the Work Function and Work Practices!* The festival in this example ends up being a success and attracts a large number of local residents, significantly more than in previous years due to the locally relevant theme as well as the specific marketing efforts of the team. Post-festival, a member of the marketing team reflects on this and decides to propose a change to the system of their work practices: would it not be better to integrate the collaborative meeting between production team and artists into the marketing workflow for subsequent years and to also invite the marketing director along to the meeting in order for them to hear first-hand what creative direction is chosen before starting the design process? This therefore becomes an institutional change with the aim of improving work practices involving different teams within the organisation.

4   *Improve Products and Services!* Finally, for the festival, future marketing efforts can then be improved through using and applying this new knowledge embedded in the discussed work functions and practices. This final step is, not surprisingly, often achieved when employees start collaborating with other members of the organisation outside their immediate team, or with other external stakeholders who might have a different perspective or way of thinking about certain tasks – in this case, the production team and headline acts who have never directly worked with the marketing team before, but rather met separately and then sent their final decision to them.

This simple example shows that overall, Wiig's (1995) model is useful in thinking more broadly about knowledge flows within an organisation and to encourage all employees to engage in making suggestions for improvement for building, pooling, distributing and applying knowledge. It is perhaps the most pragmatic of all knowledge management models, but a major downside of this model is that to date there is a lack of further research as well as practical implementation of it (Dalkir, 2017).

McElroy's knowledge management cycle

McElroy (2003) was among the first to take the basic assumptions of knowledge management a step further and argued that while first generation knowledge management practitioners assumed that valuable knowledge already exists in organisations, second generation knowledge management rests on the notion that knowledge needs to be produced in human social systems, through individual as well as shared processes. Based on this assumption, knowledge management is therefore not only about the integration of existing knowledge, but more about the production of knowledge in the first place.

His view is centred around the following key principles and assumptions (for a more in-depth discussion of these and an illustration of the framework, see McElroy, 2003: 7-9):

♦   As people experience a problem or gap to something they are working on, they tend to engage in learning in order to be able to take actions and achieve the desired outcome.

- This process of learning eventually enables employees to formulate 'knowledge claims.' Knowledge claims can be assertions or arguments, in some cases theories that describe which actions might lead to the desired outcome and therefore help solve the problem.

- The learning process can happen at the individual level but can also attract other employees to then collectively share ideas (formally or informally) and evaluate different options. At an individual and group level, ideas can be put into practice and action can be taken. Whereas, at an organisational level, some further validation needs to happen (e.g. through further evaluation by the executive team). Together the processes of formulating and evaluating different knowledge claims, in McElroy's terms, are called 'knowledge production'.

- It is important to note, that knowledge claims formulated at the individual and group level will not necessarily succeed at the organisational level. Accordingly, all knowledge claims can be classified as either 'surviving knowledge claims', if successful; or 'undecided knowledge claims' or 'falsified knowledge claims' in the case of unsuccessful ones. At an organisational level it is therefore possible to produce claims about these claims, or 'meta-claims'.

- If the success (or indeed, failure) and content and evaluation of different knowledge claims is shared with other members of the organisation, they can potentially be integrated into wider business and operational processes. This is what McElroy calls 'Knowledge Integration' (i.e., sharing and diffusing knowledge).

- Once knowledge has been integrated throughout the organisation, it will either be held by individuals or groups (agents) and take the form of beliefs or mental models, or it will be held explicitly (encoded) in artefacts, such as documents, files, databases. In other words, some of it will be in tacit form, some of it in explicit form, and some of it will be classified as 'subjective knowledge', some of it as 'objective knowledge.' The sum of all these knowledge claims held within an organisation, is what McElroy calls the 'Distributed Organisational Knowledge Base' (DOKB).

To illustrate this complex process, an event-related example will be discussed: an event company is tasked with producing a high-profile awards ceremony for a client, including a VIP reception for 50 guests in a separate room, followed by the actual ceremony with 600 guests in the main auditorium. The date, time, theme, venue, and guest list have been confirmed. However, the client then decides to increase the VIP list from 50 to 80 guests, which means the event team are faced with a problem first and foremost in terms of venue capacity, as well as additional cost for the VIP reception. The production team start formulating 'knowledge claims' – they identify different actions which might lead to the desired outcome and therefore help solve the problem. The team comes to the conclusion that one option is to move the VIP reception to a different area of the venue (the foyer area, rather than a separate room, which will be blocked off and entry will be for VIP ticket holders only to avoid the general public to access this area). Money saved from not having to rent an extra room can then be spent on additional food and drinks for the VIP reception to cater for the additional 30 people. The second option would be to change the venue altogether and find a new sponsor to cover the extra costs associated with that. Both options are evaluated, and the team decide to go with option 1, then present this to the executive team for further validation. According to McElroy, new knowledge has now been produced.

In a second step, this new knowledge is then being integrated (shared with others, diffused throughout different teams) into the wider organisation: the technical team need to make the necessary adjustments, the communications team redesign and print VIP invites to include tickets for the VIP reception, and the catering supplier is being informed about extra numbers. Some of the new knowledge is therefore held by individuals and groups (tacit knowledge, 'how things are done'), some of it is made explicit and stored in artefacts, such as email exchanges between production team and catering, meeting minutes, and the newly updated VIP database. Together these different knowledge claims form the Distributed Organisational Knowledge Base, which can then be used and applied in the business process environment (i.e., put into practice as the event unfolds). At this stage, yet new problems might arise, which will trigger the entire process to start again.

It should be noted that within his Knowledge Life Cycle framework, McElroy (2003: 9) also includes Business Processes and highlights that the Knowledge Processing Environment and the Business Processing

Environment are connected and constantly interact with each other. This is where the entire process gets very complex, because it, "begins with the detection of problems by agents in the context of business processing […], and ends with the choice of newly validated knowledge claims, beliefs, and belief predispositions in the DOKB and its containers. Knowledge use, which later follows, occurs within the context of business processing, not knowledge processing, and it is in the midst of knowledge use in business processing, in turn, that new problems arise and are detected." It is therefore a never-ending process.

McElroy's (2003) complex framework has been cited a lot in the knowledge management literature, but nevertheless, a few issues with it need to be pointed out: Loan (2006) highlights that to define knowledge as a product is a limiting assumption and the opportunities for implementing the model for organisational learning are therefore also limited. The framework is also underpinned by a range of other theories, yet its usefulness in conjunction with each of these is not clear (e.g., claims about knowledge, truth and wisdom; knowledge management and sustainable innovation). It is, however, still a valuable tool in thinking about knowledge production and integration processes within the wider business processing environment.

Objectivist vs. practice-based perspective on knowledge management

Knowledge activities, models and frameworks presented in this chapter represent one type of knowledge management approach, which Hislop et al. (2018) called the 'objectivist perspective' on knowledge and knowledge management. It is based on the understanding that knowledge is an object and it can be separated from those who possess it. However, some of the models discussed in this chapter are also underpinned by what Hislop et al. (2018) refer to as the 'practice-based perspective' on knowledge. This understanding of knowledge and knowledge management started to develop in the 2000s and will be covered more in-depth in the next chapter. It highlights the development of knowledge practices (Chapter 4), which further enhance the activities and processes covered in this chapter, but acknowledges that 'knowledge' in itself cannot be 'managed', because it cannot be separated from the people who possess it. Rather, organisations need to create an organisational structure (Chapter 5) and organisational culture (Chapter 6) that support knowledge creation, utilisation, transfer, and documentation practices.

Study and discussion questions

- Which of the three dimensions of intellectual capital (human, organisational, social capital) is the most important one to consider in an event organisation and why?
- In your own words, outline the six main knowledge activities and provide an event-related example for each.
- Define 'knowledge sharing/transfer' and highlight challenges for transferring different types of knowledge between event stakeholders.
- Which one of Nonaka and Takeuchi's (1995) knowledge conversion stages (socialisation, externalisation, combination, internalisation) would you say is the most difficult one to achieve and why? Identify event-related examples at the individual, group and organisational levels.
- Identify the importance as well as limitations of using ICT during each of the stages of Wiig's (1995) knowledge management model.

Recommended additional readings

Blackman, D., Benson, A. M., & Dickson, T. J. (2017). Enabling event volunteer legacies: A knowledge management perspective. *Event Management*, 21(3), 233-250.

Heisig, P. (2009). Harmonisation of knowledge management – comparing 160 KM frameworks around the globe. *Journal of Knowledge Management*, 13(4), 4-31.

Nonaka, I., & Konno, N. (1998). The concept of 'Ba': Building a foundation for knowledge creation. *California Management Review*, 40(3), 40-54.

References

Alavi, M., & Leidner, D. (2001). Review: Knowledge management and knowledge management systems: Conceptual foundations and research issues. *MIS Quarterly*, 25(1), 107-136.

Alvesson, M., & Kaerreman, D. (2001). Odd couple: Making sense of the curious concept of knowledge management. *Journal of Management Studies*, 38(7), 995-1018.

Blackman, D., Benson, A. M., & Dickson, T. J. (2017). Enabling event volunteer legacies: A knowledge management perspective. *Event Management*, 21(3), 233-250.

Clayton, D. (2016). Volunteers' knowledge activities at UK music festivals: a hermeneutic-phenomenological exploration of individuals' experiences. *Journal of Knowledge Management*, 20(1), 162-180.

Dalkir, K. (2017). *Knowledge Management in Theory and Practice* (3rd ed.). Cambridge, MA: MIT press.

Getz, D. (2018). *Event Evaluation*. Oxford: Goodfellow Publishers.

Halawi, L., McCarthy, R., & Aronson, J. (2017). Success stories in knowledge management systems. *Issues in Information Systems*, 18(1), 64.

Halbwirth, S., & Toohey, K. (2001). The Olympic Games and knowledge management: A case study of the Sydney organising committee of the Olympic Games. *European Sport Management Quarterly*, 1(2), 91-111.

Heisig, P. (2009). Harmonisation of knowledge management – comparing 160 KM frameworks around the globe. *Journal of Knowledge Management*, 13(4), 4-31.

Heisig, P. (2014). *Knowledge Management - Advancements and Future Research Needs*. Paper presented at the British Academy Of Management Conference, Belfast.

Hislop, D., Bosua, R., & Helms, R. (2018). *Knowledge Management in Organizations - A Critical Introduction* (4th ed.). Oxford: Oxford University Press.

Loan, P. (2006). Review of *The New Knowledge Management: Complexity, Learning and Sustainable Innovation* by Mark McElroy. *On The Horizon*, 14(3), 130-138.

McElroy, M. W. (2003). *The New Knowledge Management – Complexity, Learning, and Sustainable Innovation*. Amsterdam: Butterworth Heinemann.

Muskat, B., & Deery, M. (2017). Knowledge transfer and organizational memory: An events perspective. *Event Management*, 21(4), 431-447.

Nonaka, I., & Takeuchi, H. (1995). *The Knowledge Creating Company – How Japanese Companies Create the Dynamics of Innovation*. New York: Oxford University Press.

Nonaka, I., & Konno, N. (1998). The concept of 'Ba': Building a foundation for knowledge creation. *California Management Review*, 40(3), 40-54.

Nonaka, I., & Toyama, R. (2003). The knowledge-creating theory revisited: knowledge creation as a synthesizing process. *Knowledge Management Research & Practice*, 1(1), 2-10.

Nonaka, I., Toyama, R., & Konno, N. (2000). SECI, Ba and leadership: a unified model of dynamic knowledge creation. *Long Range Planning*, 33, 5-34.

Ragsdell, G., & Jepson, A. S. (2014). Knowledge sharing: insights from Campaign for Real Ale (CAMRA) festival volunteers. *International Journal of Event and Festival Management*, 5(3), 279-296.

Singh, N., & Hu, C. (2008). Understanding strategic alignment for destination marketing and the 2004 Athens Olympic Games: Implications from extracted tacit knowledge. *Tourism Management*, 29(5), 929-939.

Singh, N., Racherla, P., & Hu, C. (2007). Knowledge mapping for safe festivals and events: An ontological approach. *Event Management*, 11(1-2), 71–80.

Stadler, R. (2019). Knowledge management in event and festival organisations: Challenges and future directions. In E. Lundberg, J. Armbrecht, & T. Andersson (Eds.), *A Research Agenda for Event Management* (pp. 154-169). Cheltenham: Edward Elgar.

Stadler, R., Fullagar, S., & Reid, S. (2014). The professionalization of festival organizations: A relational approach to knowledge management. *Event Management*, 18(1), 39-52.

Toohey, K., & Halbwirth, S. (2005). *Sport Event Management and Knowledge Management: a useful partnership*. Paper presented at the Impact of Events Conference. Proceedings of the 2005 Events Management Research Conference, Sydney.

Tow, W. N.-F. H., Venable, J. R., & Dell, P. (2015). *Developing a Theory of Knowledge Identification Effectiveness in Knowledge Management*. Paper presented at the PACIS.

Turulja, L., & Bajgoric, N. (2018). Information technology, knowledge management and human resource management. *VINE Journal of Information and Knowledge Management Systems*, 48(2), 255-276.

Wiig, K. M. (1995). *Knowledge Management Methods - Practical Approaches to Managing Knowledge*. Arlington, TX: Schema Press, Ltd.

Wiig, K. M. (2004). *People-Focused Knowledge Management – How Effective Decision Making Leads to Corporate Success*. Amsterdam: Elsevier Butterworth Heinemann.

4 Relational and Practice-Based Knowledge Management

Learning objectives

- ☐ Understand the difference between the objectivist and the practice-based perspective on knowledge management.
- ☐ Define 'knowing' and 'know how'.
- ☐ Discuss relational, embedded and embodied knowledge and the role emotions play in practising knowledge management.
- ☐ Understand the importance of both formal and informal organisational rituals for effectively practising knowledge management.
- ☐ Explore communities-of-practice theory and apply it to event examples.

Introduction

As mentioned in Chapter 3, knowledge is by many seen as an entity or an object that can be possessed by people but can also exist completely independently of people. This objectivist perspective on knowledge, however, has over the years been critiqued a lot, and a different approach to knowledge management, or even a different understanding of knowledge itself has emerged: knowledge, or as some prefer to say – 'knowing' or 'know-how', is now regarded as a practice and it is therefore inseparable from human beings (Gherardi, 2000; Orlikowski, 2002; Hislop et al., 2018). Hislop et al. (2018) refer to this as the practice-based

4: Relational and Practice-Based Knowledge Management 71

perspective on knowledge, whereby engaging in practices means that people do not just engage in cognitive processes, but in more holistic processes involving the whole body. It is based on the assumption that knowledge is not an object, but rather it is multi-faceted and complex, explicit and tacit at the same time, individual and distributed, situated and abstract, mental and physical, static and constantly developing and evolving (Blackler, 1995). Knowledge in itself therefore cannot be managed; rather, the management of knowledge can to some extent be supported and facilitated by collaboration and interpersonal communication. Hislop et al. (2018: 42) summarise a number of ways of how this can be achieved, some of which will be discussed in this and the following two chapters:

♦ Developing a knowledge sharing culture (through rewarding people for sharing);

♦ Facilitating the development of organisational communities-of-practice;

♦ Providing forums (electronic or face-to-face) which create opportunities for social interaction between people;

♦ Implementing a formalised 'mentoring' system to pair experienced and inexperienced workers;

♦ Designing job roles to facilitate and encourage inter-personal communication and collective problem-solving.

This chapter summarises and discusses the practice-based perspective on knowledge management, including questions around knowing and know-how, differences between embedded and embodied knowledge, formal and informal knowledge ritual practices, and the role that emotions can play in knowledge management. It also introduces the idea of communities-of-practice as one specific approach to knowledge management that is in line with this practice-based understanding of knowledge. It is important to note that, while ICT was a key element in the knowledge activities discussed in Chapter 3, according to the practice-based perspective on knowledge, ICT-based repositories are not very useful for the storage and documentation of knowledge, as critical (tacit) elements of knowledge will always be missing even once the knowledge has been codified. Some of the ICT-supported activities in Chapter 3 can therefore potentially enhance certain knowledge practices (e.g. to facilitate communication, or to identify experts within the organisation), but it will always require people interacting and engaging in practices to actually produce, create, use, share, and store knowledge.

'Knowing', 'know-how', embedded and embodied knowledge

Knowledge as understood in the practice-based approach to knowledge management, can better be described as 'knowing' or 'know-how'. This means that know-how is always socially constructed, and therefore subjective and open to interpretation. It does not simply exist out there. Knowing "involves the active agency of people making decisions in light of the specific circumstances in which they find themselves in" (Hislop et al., 2018: 36). Or, in other words, producing new knowledge therefore requires people to actively engage in a process of constructing meaning. But because this process is very much based on people's interpretation and understanding of various things, knowledge will never be unbiased or completely objective; it is always shaped by the values and culture of those who produce it (Hislop et al., 2018).

This relational understanding of knowledge management or 'knowing in practice' was first introduced by Gherardi (2000) and Orlikowski (2002). They argued that the 'know-how' (rather than merely the 'know-what') is most important to individuals as well as organisations as a whole, because it is based on tacit knowledge and the specific 'ways of doing things' within an organisation (Orlikowski, 2002; Clegg & Ray, 2003; Kellogg et al., 2006). Knowing is a social practice, it is enacted through and embedded in people's everyday activities. It does not simply,"exist 'out there' (incorporated in external objects, routines, or systems) or 'in here' (inscribed in human brains, bodies, or communities)" (Orlikowski, 2002: 252). Burr (2003: 9) argued that, "[k]nowledge is therefore seen not as something that a person has or doesn't have, but as something that people do together." One could further say that knowledge itself is on the one hand embedded in work practices, tasks and routines, on the other hand it is also embodied by the people who carry out and engage in these practices. This means that people's knowledge can only develop when they conduct activities, engage in practices and therefore gain experiences (Hislop et al., 2018). Hence, we should not aim to define and make knowledge itself explicit, but rather try to understand the context, relations and practices through which knowledge is produced, enacted, embodied and shared (Orlikowski, 2002; Yahya & Goh, 2002; Feldman & Orlikowski, 2011; Stadler, 2019). Individual employees can share these practices over time and even change them using their existing skills and knowledge, which in turn creates new knowledge for the organisation as a whole. As summarised by Corradi et al. (2010), the practice-based perspective on knowledge management acknowledges both the historical and structural context in which actions take place. It combines knowledge management theory with activity-theory, actor-network theory and situated learning theory.

An example of practice-based and relational knowledge management would be mentoring approaches: a lot of event and festival organisations advocate mentoring approaches to sharing knowledge. Based on the understanding covered in this chapter, the mentor here would be regarded as someone who possesses a lot of knowledge on key tasks, some of which is embodied knowledge which cannot be made explicit and simply passed on to the mentee. It requires the mentee to not only communicate with the mentor over an extended period of time, but to also interact and work with them in different ways, so that over time the mentee can start to understand the knowledge embedded in different work practices and embody some of it themselves through practice and experience. In other words, acquiring and sharing knowledge can only occur through two processes: 1) immersion in practice, and 2) rich social interactions with other people. Learning by doing is probably the most effective form of achieving this, as it involves both immersion in the practice itself, as well as social interaction with others. In an event organisation, this can for example be achieved through a new staff member attending and observing a production or a client meeting, or through a final dress rehearsal for the technical team. A lot of knowledge practices, however, are taken for granted and invisible to staff members themselves, especially in an environment where there is not enough time to sit back and reflect on these practices (Stadler & Fullagar, 2016).

Formal and informal knowledge practice rituals

When knowledge practices become more and more routinised within an organisation, they can over time develop into rituals for creating or sharing knowledge. Rituals not only "(...) reflect and reinforce belief systems" (Samier, 1997: 419), but they also reveal some of the taken for granted and tacit elements of 'how things are done' within the organisation (Schein, 2004; Islam & Zyphur, 2009; Smith & Stewart, 2011). By sharing rituals, staff members learn 'how to' contribute to the organisation's goals and 'how to' collaborate – valuable tacit knowledge and know-how that needs to be acquired over time. In any organisation both formal as well as informal rituals exist ranging from "(...) 'full' or complete rituals to 'ritual-like' activities" (Smith & Stewart, 2011: 114). In formal meetings, for example, information and knowledge about 'what' needs to be done can be shared. Informal rituals on the other hand can help employees develop an understanding of 'how' these tasks will be performed and achieved. Quite often, formal rituals are associated with creating, sharing and storing the more explicit type of knowledge, whereas informal rituals help develop and share the more tacit knowledge. However, both types of rituals are in reality crucial for the transfer of *both* explicit and tacit knowledge (Armistead & Meakins, 2002) and they can be considered a specific knowledge practice.

Formal rituals are "frequently repeated, in a form largely laid down in advance" (Visser, 1991: 18) and they are part of any organisation; they not only convey shared meaning about what the organisation is and aims to achieve but also provide a platform for knowledge to be created and shared (Cabrera & Cabrera, 2005). In formal rituals, such as staff meetings, team meetings, annual celebrations or similar, members of the organisation embody, share and reinforce the organisation's values and culture, and therefore some of their practical and relational knowledge (Eisenberg et al., 2010). At the same time formal rituals emphasise staff members' 'know-how' of contributing to the organisation. Formal rituals therefore provide opportunities for knowledge to be practised and shared both internally as well as with the general public such as in ceremonies (Rowlinson et al., 2010). Different formal rituals can serve different purposes (Islam & Zyphur, 2009), for instance, full staff meetings are usually different to group or team meetings. An example of a team meeting and its associated rituals can be seen in Case Study 1.

In practice, as well as in academic literature, a lot of emphasis has been put on creating formal information and knowledge-sharing rituals through meetings (Smith & Stewart, 2011). However, informal rituals and even just 'ritual-like activities' (Smith & Stewart, 2011: 114) are equally as important in organisations and can take on many different forms, such as having lunch or coffee together. For example, when employees from different work teams are having lunch together, they can share similarities and differences in their approaches to work tasks and resolve common issues faced, while generating mutual trust and support through engaging in a pleasurable activity: eating (Stadler & Fullagar, 2016). These informal rituals therefore provide opportunities for sharing knowledge on what employees are currently working on and how they are performing their tasks, as well as for making sense and interpreting the 'how to' of the more formal information shared in meetings (Smith & Stewart, 2011).

Informal rituals can therefore help create and enhance the social relationships and trust necessary for effectively practising knowledge creation and transfer. They are crucial within this practice-based and relational understanding of knowledge management. Informal rituals such as eating, drinking together or talking in the corridor can also bring about shared values and beliefs and symbolise community (Trice & Beyer, 1984). Dixon (1999: 47, emphasis in original) highlighted that, "[h]allways are places where *collective meaning* is made—in other words, meaning is not just exchanged, it is *constructed* in the dialogue between organizational members." In turn, such informal rituals constitute staff members' 'know-how' that enables collaboration and effective practices of knowledge transfer. Informal, spontaneous conversations and rituals are very valuable in terms of knowledge sharing, as not only good news can be shared, but staff members are also more receptive to bad news. However, the importance of these informal practices is oftentimes overlooked in the high-pressure and stressful event work environment. Ideally, a combination of both, formal and informal rituals should be aimed for in an organisation, in order for knowledge to be effectively practised in these two different, but mutually reinforcing ways.

Case Study 1: A knowledge sharing ritual at the Queensland Music Festival, Australia

The Queensland Music Festival, originally named the Brisbane Biennial Festival of Music, was established in 1990. It is a biennial music festival, taking place in Brisbane and regional communities all over the state of Queensland, Australia. The festival includes a variety of musical styles; local, national and international artists; and at the same time, encourages participation within the communities. Most events are free and accessible to all. Furthermore, the rich diversity of musical styles in Queensland is celebrated, creating identity for remote regions within the state. A lot of the artistic projects run in the communities are long-term collaborations that tell local stories and define local culture and aim to give back to the community. In 2011, I joined the festival organisation as an ethnographic researcher. I attended a number of (formal) staff and team meetings as part of this, which provided opportunities for knowledge to be created and shared, and for team members (as well as myself) to come to know 'how to' work together and collaborate as well as learn 'how to' perform their roles within the team. A risk assessment meeting will be described below where the technical director met with a technical manager and a contractor to discuss the specific risks associated with each QMF project. The meeting represents how staff members exchanged explicit, as well as tacit knowledge in meetings, and how they worked together and collaborated.

Field notes from this particular meeting on 17 June 2011 are presented below, as they are important to demonstrate the entire ritual. In italics, an analysis and interpretation of how the team members engaged in different processes and practices of knowledge transfer has been added, highlighting the importance of the ritual in this (all names have been changed):

Today Mark has invited me to join him and the tech guys for the risk assessment meeting. I arrive at the office at 9:00 am and join Mark, Andy and an event security expert, Rob, in the meeting room. We sit at a round table, with a laptop in the middle and a big screen on the wall. After brief introductions, we start the meeting. Mark is in charge at first *[as the only permanent staff member with QMF at this meeting, his role at the meeting is clear. It is part of the ritual for him to officially start the meeting, introduce the agenda and make sure everybody understands their role in this.]* He explains that they have a generic risk assessment plan for the entire festival that covers all major issues that are relevant to all the projects. Then there will be an individual risk assessment for each one of the projects where certain specific elements and aspects can be added. For example, the issue of using fire on stage is only relevant for one performance, not for any other project; it does not have to be included in every single document. The documents are too long anyway, and Mark says nobody wants to read extra information that is not relevant for their particular project. Furthermore, with some venues, there are existing documents in place, which QMF can simply defer to, so this needs to be factored in as well. But, of course, that also needs to be mentioned somewhere and clarified with the venues. So today's meeting is all about specifying the individual needs for each of the projects. *[Mark's brief introduction of what the meeting is about gives the team an overview and idea of what needs to be discussed and what can be left out or dealt with individually. It can be classified as a formal ritual up to this point.]*

Andy runs us through the projects he is responsible for. He has all the details about the venues, size, emergency exits, crowd control, etc. at hand. He is very well prepared as usual. *[Individual information and knowledge is now shared.]* Mark is in the meantime working on the overall document. The other two check back with him at times, as he is working on the big screen so we can all see the changes he makes. *[Simultaneously, some of the information is made explicit right on the spot. Therefore, not only is tacit knowledge converted into explicit knowledge, but this explicit knowledge is immediately shared between the three of them, as is frequently the case in formal rituals, such as team meetings.]*

Rob is very focused too; he asks tricky questions and wants to know all the details. All three of them are very knowledgeable about potential risks and issues. *[Even though they are all highly experienced, only together can they go through all the details and make sure everything is covered. Their roles are clearly distributed: Mark knows the broader scheme of how each individual project fits into the whole QMF program, Andy has detailed information about each project he is working on, and Rob brings in an outsider yet highly experienced risk management perspective.]* (...) Rob looks at Mark and says, "your mind is racing, isn't it? I can literally SEE you thinking!" Yes,

> Mark seems very focused. *[This casual statement makes everyone aware of how complex the issue really is and adds a more informal touch to the meeting. It shows how the three of them have learned 'how to' work together over time, and how not every element of a ritual needs to be formal. These informal conversations throughout the meeting are equally as important in making sense of the situation, reinforcing ways of working together, and the more tacit knowledge required to perform the task.]*
>
> It is complicated, the three guys struggle to find the right words to clearly express what they mean, go back and forth, discuss, assess and reassess, and explain. *[The complexity of the meeting is apparent; data and information is put together and shared, and collective meaning is created around it. The group is engaged in various practices of knowledge creation and transfer.]*
>
> (Based on research by Stadler, 2013)

The role of emotions in knowledge practices

When looking at the practice-based and relational understanding of knowledge management, the role of emotions also needs to be considered, as emotions underpin ways of behaving, as well as practices of creating and sharing knowledge with others. Within any organisation, certain forms of emotional expression are acceptable whereas others need to be hidden or suppressed. Emotions such as anxiety, distress or fear, for example, may not be productive in terms of knowledge management, as they can limit employees' interest in seeking information and co-creating knowledge. Emotions such as shame, guilt or sadness also build a negative context for learning (Choo, 2006). Especially in terms of learning and creating new knowledge, emotions can hence be an enhancing and/or an inhibiting factor (Hislop et al., 2018). Some employees might feel very positive emotions when it comes to discovering new ideas and knowledge or developing new and improved ways of working; whereas others may feel anxious about letting go of something familiar and dealing with some level of uncertainty. In the most extreme sense, it might mean that an employee has to completely give up everything that makes them feel safe, secure, competent and knowledgeable. On the other hand, learning from mistakes can be an emotional experience too where one can feel embarrassed, overwhelmed or powerless. Any work practice can therefore lead to conflicting emotions, where an employee for example, feels excited and anxious at the same time. This can in turn have an impact on the way they create and share knowledge with others and needs to be taken into account when discussing knowledge practice theory and in particular, relational knowledge practices.

Furthermore, in event organisations, emotions are ever-changing throughout the event life cycle. Knowing 'how to' conduct oneself, 'how to' work with others and 'how to' manage one's emotions within a certain organisational culture, however, often becomes taken for granted and thus constitutes a tacit knowledge practice. Rituals, such as the ones discussed above, and informal conversations can convey information about what kind of emotions may be expressed within the organisation and to whom (Domagalski, 1999). In other words, knowledge management constituted and practised in and through rituals necessarily includes managing one's own as well as other people's emotions throughout the process. Dealing with many different tasks and different stakeholders at the same time, long hours, fatigue and burnout, can however easily turn the event environment into an emotional pressure cooker. Acknowledging and making explicit these emotions is hence crucial for staff members in order to effectively work together and collaborate. Understanding how emotions shape certain organisational dynamics is an important element of relational knowledge practices and of working professionally (Gergen, 1999; Vince & Gabriel, 2011). This can be particularly difficult in event organisations, where time is limited to sit back, reflect and process these emotions.

Communities-of-practice

The remainder of this chapter will discuss communities-of-practice as a specific example of the practice-based understanding of knowledge management. The concept of communities-of-practice was first introduced by Lave and Wenger (1991), who referred to a certain process of social learning that occurs when people have a common interest in a subject. They share ideas, find solutions, and build innovations over time. In other words, communities-of-practice are, "a group of people who have a particular activity in common, and as a consequence have some common knowledge, a sense of common identity, and some element of a shared language and overlapping values" (Hislop et al., 2018:

196). It is the three elements of 1) common knowledge, 2) overlapping values, and 3) shared identity, which allow the formation of a shared understanding of values and assumptions within a community-of-practice, as well as create social conditions which enhance knowledge sharing, creation and utilisation.

Wenger (1998: 6-7) later applied the concept of organisational learning to communities-of-practice and gave it a new meaning within the theory of knowledge management. He argues that "communities-of-practice are everywhere. […] They are so informal and so pervasive that they rarely come into explicit focus, but for the same reasons they are quite familiar." In a way, communities-of-practice are like work teams, that make it easier to share information and work among their members, to allow all members to contribute and share their expertise, and to effectively deal with conflict and interpersonal problems. They are different to business units or departments though, in that they are evolving and internally negotiated. They are self-managed rather than defined by the organisation, and therefore very informal. They have been applied to various kinds of organisational settings (Roberts, 2006), including festival organisations (Abfalter et al., 2012). For a summary of some of the main differences between formal work teams and communities-of-practice, see Table 4.1. Participation, situated learning, bonding, and creativity and innovation are among the key features of communities-of-practice and will be discussed in relation to knowledge management below.

Table 4.1: Differences between formal work groups and communities-of-practice.

Work group or team	Community-of-practice
Formally and externally defined	Informal, evolving and internally negotiated, membership is voluntary
Structured around a specific service and/or product	Members share a collective practice or knowledge
Formalised relations	Informal, inter-personal relations
Usually hierarchical structure	Non-hierarchical, fluid, self-managing
Permanent, or with pre-set timeframe/objective	Indefinite, no set timeframe, forms and dissolves whenever needed

Participation

Participation is perhaps the most important element of creating a good learning environment in communities-of-practice. It is through participation that newcomers become part of the community. They start from an outside perspective, from peripherality, they learn from old-timers, and through this process, they become more and more aligned with the practices of the community. This is what Lave and Wenger (1991: 29) called 'legitimate peripheral participation'. As newcomers join, the lifecycle of a community-of-practice starts. Newcomers interact with old-timers, they learn the practices, and finally become 'insiders'. They then become old-timers themselves, and yet another set of newcomers come in and are taught by the old-timers, until the old-timers are finally replaced. This process is ongoing and continually changing. Every member of the community has their duties, newcomers as well as old-timers, and these duties change over the lifetime of a community and make the community go through various stages. The active process is important, because newcomers want to develop their own view of the organisation (Brown & Duguid, 1991; Lave & Wenger, 1991). In an event organisation staff members continuously go through this process – new seasonal staff and volunteers join at different times, participate at different levels, and become 'insiders.'

In terms of knowledge management, through participation, explicit, formal and systematic knowledge is shared, and converted into tacit knowledge. Wenger (1998: 58) called this 'reification', which describes making sense, giving a name to something, interpreting, using, or making meaning. Thus, it is through participation that people learn something and through reification that they internalise it and give it a meaning. Participation and reification cannot exist on their own; one brings about the other, much as 'knowing' and 'doing' are also inseparable based on this framework, and in line with the practice-based understanding of knowledge management. Participation can further happen at three different levels (see Figure 4.1): the *core group*, the *active group*, and *peripheral members* (Wenger et al., 2002):

♦ The *core* group is the most important part of the community. It is comprised of a small group of people, who meet and discuss important topics on a regular basis. They can be seen as the leaders of the community.

♦ The *active* group is further outside the core, but members of this group are still very much involved in the community in that they attend meetings, for example.

♦ A lot of community members are *peripheral*. They do not participate regularly and seem very passive. But for communities-of-practice they can indeed play a key role. They view the community from another perspective and give different insights into certain issues.

Furthermore, outside these three levels of community participation are the so-called *outsiders*. They do not belong to the community, but they have a certain interest in its activities. It is important to also consult these people on important issues and to apply their knowledge to processes within the community (Wenger et al., 2002).

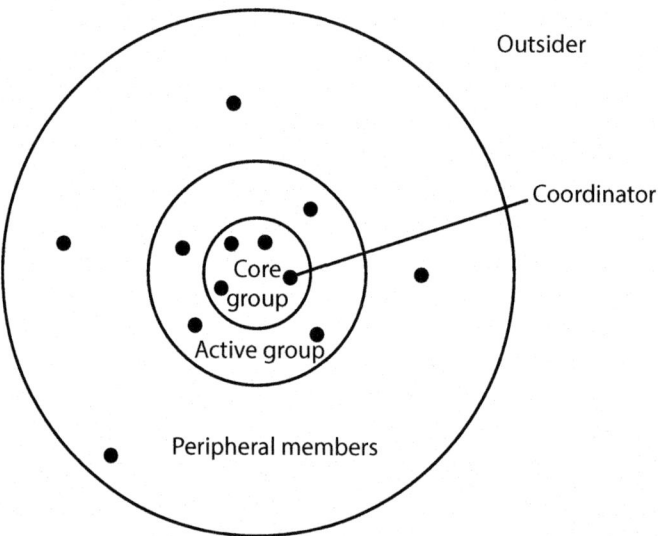

Figure 4.1: Degrees of community participation. (Wenger et al., 2002: 57), reproduced with permission from Harvard Business School Publishing Corporation

It is important to mention that members of the community-of-practice can change their degree of participation over time. There are different forms of 'trajectories': Members can move from outside the community inwards, or from the inside further outwards. Through these trajectories, people learn different views and can incorporate past knowledge into their present place within the community. At the same time, they are taught new things, told different stories, and tested by others, and can find out what the expectations of their new environment are.

They might need to change their practices and adapt to the new context, but at the same time they can share their existing knowledge with new colleagues (Wenger, 1998; Argote & Ingram, 2000). In this sense, communities-of-practice are very dynamic and constantly change and evolve (Hislop et al., 2018), which makes them interesting to investigate in an events context, where the organisational structure is 'pulsating' and therefore very dynamic as well. The different levels of participation within a community-of-practice can indeed be applied to event organisations and an example of this is provided in Case Study 2 below.

Situated learning

Lave and Wenger (1991) were among the first to argue that learning can only take place through activity and participation. This is different from conventional theories of learning that suggest that learners are passive recipients of information and knowledge, and where learning is seen more like a top-down approach, with training playing an important role. Learning is seen here as a permanent exchange of ideas and viewpoints and a giving and taking. This takes place in a social world constructed by the participants. It involves people and members of the organisation or community as well as people outside the organisation, and represents an application of the model of the three degrees of participation as described above. All members of the community-of-practice interact on an ongoing basis and put their attention to continually changing environments and situations. From this point of view, learning entails becoming an insider (Brown & Duguid, 1991; Lave & Wenger, 1991). It is in line with the practice-based understanding of knowledge management, in that learning is regarded as social and relational.

The common practices in a community do not normally develop within a short period of time. Rather they evolve over time and need continuous mutual engagement and participation in order to fully establish a common understanding and to act as learning tools. Through this evolution over time, learning also becomes the basis for communities-of-practice. Learning helps establish practices, and practices are the outcome of learning. Thus, the two always interact. Only after a certain amount of learning has been achieved, can communities-of-practice evolve. Then they develop and change over time again and again, until they finally (maybe) dissolve. They do not have a clear beginning and end; rather they develop when there is a need for a shared practice (Wenger, 1998; Handley et al., 2006).

Bonding through common problems and a common purpose

It has already been outlined that in a community-of-practice people share a common problem and work on a common issue. This is what brings them together and what creates important bonds between community members. Thus, in a community-of-practice every member has to engage with the others and share a common purpose. Strong bonds between the participants can be developed and they help members feel as an important part of the community (Brown & Duguid, 1991; Wenger, 1998). To describe the bonds and relations within a community-of-practice, Wenger et al. (2002) used three elements as the basis: a *domain*, a *community* of people, and a *practice* that these people share. These three elements can help explain communities-of-practice as a structure for creating and sharing knowledge:

- The *domain* entails all important issues and problems that members of a community-of-practice experience. It is not a written document of the problems, but rather some informal understanding of them, that evolves and changes over time. The domain is the basis for questions to be addressed and for knowledge to be created and shared. It helps to find out what is of real importance and what is just an average idea or issue. Every single member of the community has his or her own view of the domain. These views may be very different, but they create some common values and norms, some kind of community culture.

- The function of the *community* is the actual purpose of why people interact regularly. Members of the community respond to issues and problems that emerge from their domain, and thus are important and relevant to them and to the community as a whole. As discussed above, the level of participation in the group varies from one member to the other, but the community is the foundation for establishing relations and bonds with other participants.

- *Practice*, the third element, provides a common ground of knowledge that is essential to community members, because it represents their views and perspectives on particular issues. It is not the work of a single person, but of the whole group and their interactions and relations. It is not a single leader that builds a practice, but a tight bond of community members that work together, discuss, and exchange ideas. Working on a common problem or around a common purpose is thus shaped by engagement and participation in the community. Every member identifies with the community and sees themselves as an important part.

In relation to events, a community-of-practice based on these three elements therefore does not usually form within a specific department or functional area, but rather across a range of different teams. It could for example include one member of the development team, one member of the finance team, one of the marketing and another one of the production team, as well as an outsider to the organisation, such as a supplier or contractor. What brings them together is a specific task or challenge they face, embedded in the *domain* that they share. Rather than trying to solve the issue within, for example, just the production team, it is more beneficial to tap into the other members' knowledge and understanding of the problem and bring together their different perspectives from a practice-based approach. Once the problem has been dealt with, the community-of-practice might turn to another task, or it might even completely dissolve.

Creativity and innovation

A successful organisation provides a foundation for creativity and innovation among its members. Especially with creative people, new ideas can develop easily, if the creative thinking process is not limited by rules or barriers. The environment should be open, so that creative community members at all levels can carry out their ideas and suggestions. In this sense, the community-of-practice is a valuable form for enhancing creativity and innovation. The community is very diverse and yet there is a sense of familiarity among its members. Therefore, their creativity can be used more efficiently. Exchanges of ideas take place on an ongoing basis and enhance creative processes within the community (Boone, 2001; Lesser & Storck, 2001), a particularly important element in event organisations, for example when it comes to event design, theming, and staging.

Sharing a practice within a community also increases innovation, because the bonds between people that arise when working on a common problem create a valuable basis for communication and focused conversations. Stories especially provide an atmosphere within a community-of-practice that supports innovation (Brown & Duguid, 1991; Wenger et al., 2002). Brown and Duguid (1991) also mentioned that innovation in a community comes about through the interaction between the community and its environment. Every process of innovation in turn has an effect on the environment. This process, however, does not have to bring a radical change; it can be a gradual movement that establishes through continuous interaction between community members and their environment. Applying the model of degrees of participation on the process of innovation thus makes sense: exchanging ideas with other members of the community on the various levels, as well as outsiders, creates a culture of incremental innovative improvements. Peripheral members bring in new ideas to the core group; the core group establishes a new concept for the active group, and so on.

Limitations and critique of communities-of-practice

As explained above, communities-of-practice provide a valuable forum for learning. However, as everyone's knowledge becomes very important to the whole community, the power of the community can also be a threat for learning. If each member only focuses on their individual knowledge, arrogance will set in, and learning will never take place. Power needs to be distributed evenly, so that responsibility is also shared among all participants, an issue that will be further discussed in Chapter 7. This can be difficult, however, because community members usually possess different levels of expertise and those with a higher level of expertise may automatically start to exercise more power. The same is true for different levels of participation. Those with full participation may end up using their power against members with only a marginal or peripheral level of participation (Boone, 2001; Roberts, 2006).

Communities-of-practice can also become locked up in their own interpretation and view of the world. This can limit the success of the community. It is thus important that there is an ongoing exchange of ideas at all levels of the community, and also between these levels (Brown & Duguid, 1998). At the same time, if community members do not associate enough within the community, they will not be able to develop relationships and trust. Without these bonds and people's dedication, the practices will remain static. On the other hand, if people within the community are too passionate about a topic, they can easily disagree with others. If one member dominates the discussion, an argument might arise. Such strong commitment can become an obstacle for newcomers to join or for the development of new ideas. A similar situation can arise when community members also participate in other communities-of-practice where there is a different world view. This may lead to a conflict between the different communities, as well as a conflict of interest for individual community members (Handley et al., 2006). However, Brown and Duguid (1998: 103) mentioned that being part of more than one community-of-practice can also be an advantage. Individuals can serve as 'translators' and bring in ideas from other communities.

Since communities-of-practice are self-selected and based on mutual engagement, it is very common for smaller groups or cliques to form, and for friendships to arise. This might threaten the community as a whole, because as strong relationships evolve individuals may stop arguing and critiquing each other. On the other hand, it is these aspects of the community that keep it alive and active. If they are lost, the processes within the community may slow down or even stop altogether. Additionally, these tight friendships can also create a barrier for newcomers to enter the community (Wenger et al., 2002). Finally, members of the community may lose interest in further developing ideas and knowledge. If, for instance, the problems do not seem relevant to them, they will not put enough effort into finding solutions and will stop participating and interacting with other community members. The same problem can also occur on the recipient's side. If the information that the recipients receive is not important to them, they will ignore it and refuse to accept it (Boone, 2001).

Communities-of-practice have been applied to a festival organisation by Abfalter et al. (2012), see Case Study 2 for an example. It is one way of interpreting knowledge practices within an organisation, although difficult to achieve in event organisations due to some of the challenges discussed in Chapter 2 (e.g. lack of time and trust among staff members, large number of volunteers).

Case Study 2: Communities-of-practice at the Colorado Music Festival, U.S.

The Colorado Music Festival (CMF), a non-profit organisation, was founded by Giora Bernstein in Boulder, Colorado in 1976 as a classical music festival. Every summer, musicians from all over the world come together for six weeks (June – August) and perform at the Chautauqua Auditorium in Boulder. The festival is mainly attended by local residents, but also attracts tourists and international visitors. The program in 2008 consisted of four concerts per week performed by the CMF Chamber Orchestra and the CMF Festival Orchestra, as well as a Young People's Concert at the beginning of the season, and a number of Educational and Outreach Programs for all ages. The festival is managed by a small group of year-round staff, and during the festival season also includes a large team of seasonal staff (production crew, sound engineer, house manager, interns and an 80-person orchestra). The team is supported by the Board of Trustees (consisting of different committees), as well as the Friends of CMF (a group of volunteers who provide fundraising and service support throughout the year, but especially during the festival season). Using the communities-of-practice framework, members of the organisation take on different levels of participation, as can be seen in Figure 4.2 below:

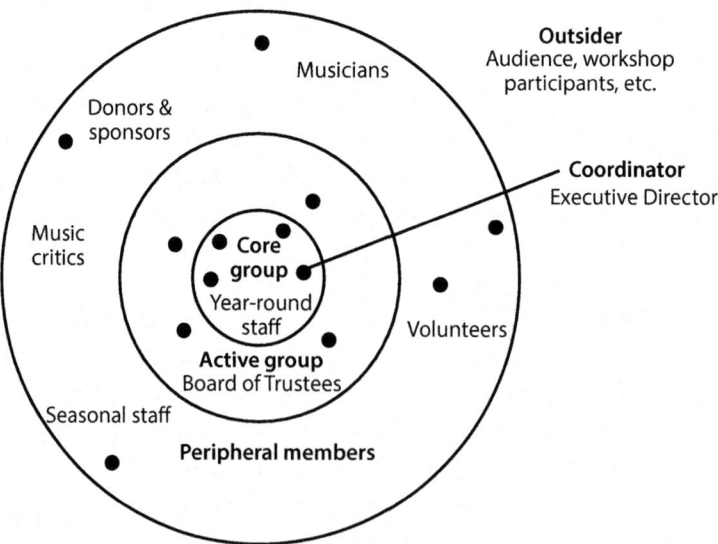

Figure 4.2: Levels of participation at CMF. (Abfalter et al., 2012: 9); Reprinted with permission from *International Journal of Arts Management*.

4: Relational and Practice-Based Knowledge Management

As can be seen here, communities-of-practice form out of this festival structure. Whenever a problem arises or staff members work on specific tasks, they arrange themselves into small communities-of-practice consisting of members at different levels of participation. These are dynamic teams; they are based on the overall organisational structure, but they form and dissolve as and when necessary. For example, in order to run a violin workshop for children aged 12-15, a member of the core group (Development Director, DD; core group) gets together with two of the musicians (two violinists; peripheral group), one seasonal staff member (production crew; peripheral group) and three volunteers (also from the peripheral group), who will help run the workshop on the day. She explains:

> DD: They [musicians, production crew, volunteers] are the key people for this workshop and strongly involved in the planning and running of it, of course I want them to be part of every meeting, every discussion we have. They bring in so much, not just knowledge and expertise, but also passion and enthusiasm for these workshops. They are absolutely key; I couldn't do it without them.

The Development Director, however, also invites along to several meetings, a member of the Board of Trustees (active group) who has a good understanding of the overall program on a more strategic level and provides valuable insights into where this workshop fits with other creative elements of the festival:

> DD: It is this bigger picture that she [member of Board of Trustees] has got, which is really useful to be reminded of every now and again. I get too busy with the everyday tasks; I sometimes can't see the wood from the trees.

Lastly, the workshop participants (outsiders: children and their parents) are also consulted along the way, in terms of what they want to get out of the workshop and who they would like to work with. Over time, some of these community members increase or decrease their levels of participation. For example, the seasonal staff member (production crew) starts off on the periphery, but becomes more and more involved in the running of the day and the different elements of it, that he ends up moving to the active and later (on the day of the workshop) even to the core group. Whereas the Development Director starts off at the core group level, but later on hands over a number of key tasks to the musicians and volunteers, and therefore she moves more towards the periphery.

> DD: Oh yes, on the day of the workshop, I just sit back and let the others get on with it. The volunteers have been fully briefed, they were involved in the process from start to finish, they are now able to do it; they don't need me anymore.

These trajectories from one level of participation to another are fluid and dynamic, and the way community members create and share knowledge is therefore also ever changing. They all need to participate in the process though, in order to create meaning together and learn specific 'ways of how things are done' within this particular community. Finally, once the workshop is over, members of the community-of-practice finish off some final tasks (thanking sponsors, gathering feedback from participants, writing a report), and then the community dissolves. Individual members of this community-of-practice move on to other tasks or engage more or less in other communities-of-practice they are part of.

(Based on research by Stadler, 2008)

Study and discussion questions

- ☐ Define the practice-based perspective on knowledge management and explain how it can be applied to event organisations.
- ☐ Why are emotions a key element of relational, practice-based knowledge management? Provide examples from an event you have organised.
- ☐ Debate whether formal or informal knowledge rituals are more valuable for an organisation?
- ☐ Explain the importance of participation in communities-of-practice and provide an example of how staff members can move from one level of participation to another from your own experience of organising an event.

Recommended additional readings

Abfalter, D., Stadler, R., & Mueller, J. (2012). The Organization of Knowledge Sharing at the Colorado Music Festival. *International Journal of Arts Management*, 14(3), 4-15.

Orlikowski, W. J. (2002). Knowing in Practice: Enacting a Collective Capability in Distributed Organizing. *Organization Science*, 13(3), 249-273.

Stadler, R., & Fullagar, S. (2016). Appreciating formal and informal knowledge transfer practices within creative festival organizations. *Journal of Knowledge Management*, 20(1), 146-161.

References

Abfalter, D., Stadler, R., & Mueller, J. (2012). The organization of knowledge sharing at the Colorado Music Festival. *International Journal of Arts Management*, 14(3), 4-15.

Argote, L., & Ingram, P. (2000). Knowledge transfer: a basis for competitive advantage in firms. *Organizational Behavior and Human Decision Processes*, 82(1), 150–169.

Armistead, C., & Meakins, M. (2002). A Framework for practising knowledge management. *Long Range Planning*, 35(1), 49-71.

Blackler, F. (1995). Knowledge, knowledge work and organizations: an overview and interpretation. *Organization Studies*, 16(6), 1021-1046.

Boone, M. E. (2001). *Managing Interactively – Executing Business Strategy, Improving Communication, and Creating a Knowledge-Sharing Culture*. MacGraw Hill.

Brown, J. S., & Duguid, P. (1991). Organizational learning and communities-of-practice: toward a unified view of working, learning, and innovation. *Organization Science*, 2(1), 40-57.

Brown, J. S., & Duguid, P. (1998). Organizing knowledge. *California Management Review*, 40(3), 90-111.

Burr, V. (2003). *Social Constructionism* (2nd ed.). London and New York: Routledge.

Cabrera, E. F., & Cabrera, A. (2005). Fostering knowledge sharing through people management practices. *International Journal of Human Resource Management*, 16(5), 720-735.

Choo, C. W. (2006). *The Knowing Organization – How Organizations Use Information to Construct Meaning, Create Knowledge, and Make Decisions* (2nd ed.). Oxford: Oxford University Press.

Clegg, S., & Ray, T. (2003). Power, rules of the game and the limits to knowledge management: Lessons from Japan and Anglo-Saxon alarms. *Prometheus*, 21(1), 23-40.

Corradi, G., Gherardi, S., & Verzelloni, L. (2010). Through the practice lens: where is the bandwagon of practice-based studies heading? *Management Learning*, 41(3), 265-283.

Dixon, N. M. (1999). *The Organizational Learning Cycle - How We Can Learn Collectively* (2nd ed.). Hampshire: Gower.

Domagalski, T. A. (1999). Emotion in organizations: Main currents. *Human Relations*, 52(6), 833-852.

Eisenberg, E. M., Goodall Jr., H. L., & Trethewey, A. (2010). *Organizational Communication - Balancing Creativity and Constraint* (6th ed.). Boston: Bedford/St. Martin's.

Feldman, M. S., & Orlikowski, W. (2011). Theorizing practice and practicing theory. *Organization Science*, 22(5), 1240-1253.

Gergen, K. J. (1999). Affect and organization in postmodern society. In S. Srivastva & D. L. Cooperrider (Eds.), *Appreciative Management and Leadership - The Power of Positive Thought and Action in Organization* (Revised) (pp. 153-174). Euclid, Ohio: Williams Custom Publishing.

Gherardi, S. (2000). Practice-based theorizing on learning and knowing in organizations. *Organization*, 7(2), 211-223.

Handley, K., Sturdy, A., Fincham, R., & Clark, T. (2006). Within and beyond communities of practice: Making sense of learning through participation, identity and practice. *Journal of Management Studies*, 43(3), 641-653.

Hislop, D., Bosua, R., & Helms, R. (2018). *Knowledge Management in Organizations - A Critical Introduction* (4th ed.). Oxford: Oxford University Press.

Islam, G., & Zyphur, M. J. (2009). Rituals in organizations: a review and expansion of current theory. *Group & Organization Management*, 34(1), 114-139.

Kellogg, K. C., Orlikowski, W., & Yates, J. (2006). Life in the trading zone: Structuring coordination across boundaries in postbureaucratic organizations. *Organization Science*, 17(1), 22-44.

Lave, J., & Wenger, E. (1991). *Situated Learning - Legitimate Peripheral Participation*. Cambridge: Cambridge University Press.

Lesser, E. L., & Storck, J. (2001). Communities of practice and organizational performance. *IBM Systems Journal*, 40(4), 831-841.

Orlikowski, W. J. (2002). Knowing in practice: Enacting a collective capability in distributed organizing. *Organization Science*, 13(3), 249-273.

Roberts, J. (2006). Limits to communities of practice. *Journal of Management Studies*, 43(3), 623-639.

Rowlinson, M., Booth, C., Clark, P., Delahaye, A., & Procter, S. (2010). Social remembering and organizational memory. *Organization Studies*, 31(1), 69-87.

Samier, E. (1997). Administrative ritual and ceremony: Social aesthetics, myth and language use in the rituals of everyday organizational life. *Educational Management Administration & Leadership*, 25(4), 417-436.

Schein, E. H. (2004). *Organizational Culture and Leadership* (3rd ed.): Jossey-Bass.

Smith, A. C. T., & Stewart, B. (2011). Organizational rituals: Features, functions and mechanisms. *International Journal of Management Reviews*, 13(2), 113-133.

Stadler, R. (2019). Knowledge management in event and festival organisations: Challenges and Future Directions. In E. Lundberg, J. Armbrecht, & T. Andersson (Eds.), *A Research Agenda for Event Management* (pp. 154-169). Cheltenham: Edward Elgar.

Stadler, R., & Fullagar, S. (2016). Appreciating formal and informal knowledge transfer practices within creative festival organizations. *Journal of Knowledge Management*, 20(1), 146-161.

Trice, H. M., & Beyer, J. M. (1984). Studying organizational cultures through rites and ceremonials. *Academy of Management Review*, 9(4), 653-669.

Vince, R., & Gabriel, Y. (2011). Organizations, learning, and emotion. In M. Easterby-Smith & M. A. Lyles (Eds.), *Handbook of Organizational Learning and Knowledge Management* (2nd ed., pp. 331-348). John Wiley & Sons.

Visser, M. (1991). *The Rituals of Dinner - The Origins, Evolution, Eccentricities, and Meaning of Table Manners*. New York: Grove Weidenfeld.

Wenger, E. (1998). *Communities of Practice - Learning, Meaning, and Identity*. Cambridge: Cambridge University Press.

Wenger, E., McDermott, R., & Snyder, W. M. (2002). *Cultivating Communities of Practice*. Boston: Harvard Business School Press.

Yahya, S., & Goh, W.-K. (2002). Managing human resources toward achieving knowledge management. *Journal of Knowledge Management*, 6(5), 457-468.

5 Structural Elements of Knowledge Management

Learning objectives

- ☐ Discuss key human resource management strategies in relation to knowledge management.
- ☐ Understand hierarchical and non-hierarchical structures for knowledge management.
- ☐ Explore the value of inter-disciplinary teams for effective knowledge management.
- ☐ Define different knowledge management roles and responsibilities within teams and organisations.

Introduction

As outlined in the previous chapters, knowledge can be practised in many different ways and technology can only to some extent support these activities. There are other structural elements that organisations can put in place though, to support and enhance knowledge practices and knowledge management. This chapter starts with a brief overview of the relationship between human resource management and knowledge management. It has been argued that the two go hand-in-hand and that effective human resource management can positively contribute to creating both a structure and a culture for knowledge management. It will therefore be covered here in relation to organisational structure and then referred back to in Chapter 6, where organisational culture will be further explored. The literature on human resource management is vast and it should be noted that only HR practices that are particularly relevant to the events industry will be covered in depth in this chapter.

Based on this introduction to HR and knowledge practices, the second part of the chapter will then explore different hierarchical and non-hierarchical structures that can enhance or inhibit knowledge management, such as top-down, bottom-up and middle-up-down knowledge management, as well as the creation of interdisciplinary teams and pods. While it is necessary to have these organisational structures in place, the structure of an organisation, however, is never fixed, but rather a dynamic constellation of relationships (Küpers, 2005). In terms of the relational and practice-based approach to knowledge management this is important to acknowledge, as it helps to understand the fluid and dynamic environment in which knowledge is practised. The chapter finishes with a discussion of specific knowledge management roles and responsibilities at different levels and for both internal and external stakeholders. These again are important to recognise in any organisation as they shape the way employees think about their roles in relation to knowledge management and therefore how they come to understand certain knowledge practices.

Human resource management and knowledge management

Research into knowledge management and human resource management suggests that the two are closely related in terms of the effectiveness of team collaboration and culture (Yahya & Goh, 2002; Currie & Kerrin, 2003; Gorelick et al., 2004). Gloet and Berrell (2003: 83) maintained that, "[t]he relationship between KM and HRM is a complex one; the more aligned the strategies underlying both of these arenas, the more contribution both can make to quality practices and overall organizational performance." By learning 'how to' work together in order to achieve the organisation's aims and objectives, staff development processes and organisational structures facilitate working together in both formal and informal ways and can thus enhance knowledge practices within the team.

Figure 5.1 represents the interrelatedness of human resource management and knowledge management. Both are highly influenced by the culture, climate, vision and values of the organisation (as will be further discussed in Chapter 6) and constantly reinforce each other. For example, selecting the right people for the job in the first place, will later have a positive impact on how they create new knowledge and use and share existing knowledge. Similarly, if staff members are motivated to do their jobs in general, they will also be more motivated to acquire new knowledge, share their knowledge and expertise with others, and effectively store it for the organisation as a whole. Together, these processes and practices can enhance teamwork, collaboration and co-creation.

Figure 5.1: Human resource management and knowledge management

The overall aim of an organisation's human resource management strategy is to have the right mix of employees in terms of number, types and skills at any given point in time to meet the organisation's present and future requirements (Nankervis et al., 2011; Lockstone-Binney et al., 2020). Not having the necessary experience and skills could easily lead to increased stress in an already stressful workplace and environment (Cranwell-Ward & Abbey, 2005). An effective HR strategy is in line with the overall organisational strategy and objectives, as well as adaptable to the external organisational environment (Mosley et al., 2001; Allen et al., 2011). The 'pulsating' nature of event organisations, as discussed in Chapter 2, however, creates particular challenges (Hanlon & Cuskelly, 2002; Lockstone-Binney et al., 2020) where "[t]he workforce builds very rapidly close to event delivery" (Van der Wagen & White, 2014: 28). The different functional areas in an event organisation conduct recruitment, selection and training processes at different points in time throughout the event life cycle in order to meet their specific staffing needs. For example, marketing professionals usually come on board several weeks or months before the event, whereas volunteers only start a few days before the event, or even just on the actual event day. In event organisations, it is also important to recruit and select employees who have already built relationships with other professionals in the industry and have learned to work with them in different contexts and for different organisations (e.g. contractors, suppliers, production companies). However, they need to be aware of the different 'ways of doing things' in different organisations, in the sense that what works in one event organisation might not necessarily work in another.

Moreover, Du Plessis (2006: 101) importantly highlighted that when recruiting staff in an organisation that is keen to enhance its knowledge management processes and practices (as indeed, any organisation should be), "the organisation [should] look for specific criteria in candidates, e.g. their orientation towards knowledge sharing and creation, their sense of openness and transparency, and their ideas on knowledge management and knowledge management culture specifically." This is often overlooked in event organisations, where the importance of effective knowledge management is still under recognised.

Nankervis et al. (2011) suggested that for the purpose of staff productivity and retention, aligning the different internal and external issues with the human resource strategy of an organisation and with the organisation's culture and values is vital. The complexity of different roles within the event can partly be overcome through employing staff members with a range of different backgrounds and experiences who complement each other in terms of knowledge and skills.

For example, instead of hiring three technical managers who all have experience in sound technology, it might be more beneficial to have one technical manager on the team who has got a lot of experience in audio, another one who covers stage and lighting, and a third one who is an expert in facilities and site building. Having inter-disciplinary rather than functional teams can be particularly beneficial here, as will be discussed further below. Furthermore, taking different personalities into account is also crucial, especially when forming teams and groups who need to work together effectively and efficiently (Hislop et al., 2018). Backgrounds, experience and personalities that complement each other are much more beneficial in terms of effectively practising knowledge management than teams and personalities that clash.

Once the right people have been recruited and selected, they can start taking on their roles within specific organisational structures, teams, pods, or similar. Even if all employees come to the job with a lot of experience and knowledge, training and developing new skills should be an ongoing process for all of them. Formal induction and training sessions, on-the-job-learning as well as mentoring practices have already been covered in previous chapters and can be highly beneficial in terms of knowledge management and organisational learning. In event organisations, general training is usually provided for all staff members in areas such as health and safety or first aid, as well as position specific training relating to their functional roles (Van der Wagen & White, 2014; Lockstone-Binney et al., 2020). The most important element of staff development, however, is on-the-job learning, through which more complex experience and tacit knowledge can be gained as well as relational knowledge practices developed. Not only can new or seasonal staff members learn from the permanent staff, but also from each other. On-the-job learning, therefore, also provides opportunities for socialising and building relationships and interpersonal skills (Werner & DeSimone, 2006; Nankervis et al., 2011). Whereas a certain amount of knowledge or a university degree is expected in the events industry today, the more tacit experiences, skills and relationships can only be built on the job (Arcodia, 2009). Staff members who have gained lots of tacit knowledge and experience become highly sought after in the industry, particularly when there is limited time to further develop skills and know-how.

Another way of developing skills and knowledge over time worth mentioning, is job rotation. Job rotation is regarded as a highly effective human resource development approach. Currie and Kerrin (2003) found in their study on human resource management in a pharmaceutical company that those employees who engaged in job rotation or lateral career movement had a better understanding of the relationships between different functions and their specific roles and responsibilities. In terms of knowledge management, therefore, through job rotation staff members can develop a broader understanding of the organisation and their know-how to contribute to the organisation's aims, as well as develop strong relationships with different individuals and teams within the organisation. If staff members later need to fill in for someone who is on leave, they already possess the required skills and embodied knowledge which can be useful in event organisations during, for example, stage set-up, rehearsal periods, or performances. In an event organisation this can be achieved for instance, through hiring a seasonal staff member in a general admin role, where they learn a little bit about everything. They can then move on to a more permanent position within the organisation (depending on the organisational structure) and use their knowledge and skills as well as broader understanding of 'how things are done' in this new role. Similarly, volunteers who demonstrate a lot of passion and enthusiasm for their jobs, as well as a willingness to learn, who move around and work for different teams, can perhaps become volunteer team leaders over time, and can potentially even be hired for paid positions in later years.

Lastly, retaining workers who possess relevant knowledge and know-how is crucial, as otherwise knowledge can leak into other organisations or be lost altogether. High staff turnover and the 'pulsating' nature of events, create specific challenges for knowledge management here. For example, if a staff member who has built a long-term working relationship with a key sponsor or partner, leaves once the event is over, their tacit knowledge, know-how, as well as the relationship with the sponsor will be lost. Or, in some cases, they could even take this sponsor with them and move on to another event organisation. To overcome these issues, recognition and reward are commonly mentioned as possible ways to motivate employees and create a sense of organisational loyalty. In other words, if they experience a sense of loyalty to the event, even volunteers and seasonal staff will then want to come back for the next event season and are committed to their roles, albeit only temporary or seasonal. Rewards can be financial or non-financial in nature (Hislop et al., 2018), for example, event staff members can be recognised for their knowledge contributions through acknowledging best practices or sharing stories of knowledge management successes. This, alongside all other human resource management activities, of course, needs to be evaluated along the way in order to determine what works and what can be improved, as well as lessons that can be learned (Getz, 2018).

Top-down, bottom-up and middle-up-down knowledge management

Top-down management, or the typical pyramid structure of management, is the classic hierarchical model found in many organisations. In terms of knowledge management, however, according to Nonaka and Takeuchi (1995), it is only really suitable for information processing: at the bottom of the pyramid, simple information is produced, which is then passed up through middle managers to the top level. At the top, senior management use this information to create operational plans which in turn are passed down the pyramid again and implemented by front-line employees. In other words, in this type of organisation, only top management are creating new knowledge, while middle managers create the operational conditions, which front-line employees implement in their day-to-day practices in mainly routine work. This approach is therefore highly functional and pragmatic but limited in terms of creating new knowledge. It can, however, be found in the events industry, where many organisations are hierarchically structured.

The bottom-up management approach, on the other hand, is essentially the opposite. Nonaka and Takeuchi (1995: 126) argued that there is a lot of autonomy in this approach to knowledge management, because knowledge is mainly created, controlled and used at the bottom of the organisation. The shape of this type of organisation, however, is rather flat and horizontal. Top managers are considered the "sponsors of entrepreneurially minded front-line employees", who largely work on their own. Front-line employees do not engage in much dialogue with other employees, neither vertically nor horizontally, and tend to create new knowledge individually rather than in groups or teams. Top management do not provide much instruction or orders, they simply let front-line employees do their tasks. This type of organisation can also be found in the events industry, for example, in organisations that are entirely run by volunteers with no formal hierarchical structure or job titles that would imply a certain hierarchy.

It is commonly argued though, that in knowledge-intensive firms and industries, such as the event industry, organisations tend to be less hierarchical, and therefore boundaries for knowledge management are a bit fuzzier (Robertson & Swan, 2004; Hislop et al., 2018). Especially when regarding event organisations as 'learning organisations', it is beneficial to have less centralised and hierarchical systems of control, and rather have a flat organisational structure, or at least limited top-down control and more autonomy in decision-making. Nonaka and Takeuchi (1995) similarly argued that neither top-down nor bottom-up approaches to management are appropriate for effective knowledge management. Instead they proposed a middle-up-down organisational structure, which puts the middle managers at the centre of knowledge creation and redefines the roles for top management as well as front-line employees: "knowledge is created by middle managers, who are often leaders of a team or task force, through a spiral conversion process involving both the top and the front-line employees (i.e., bottom). The process puts middle managers at the very center of knowledge management, positioning them at the intersection of the vertical and horizontal flows of information within the company" (Nonaka & Takeuchi, 1995: 127).

In other words, the middle managers can be empowered in their key knowledge management position, and they can therefore be considered the bridge or knot between top management and front-line employees. While the top managers are working on visions, ideas and strategies, and the bottom-level employees are busy with day-to-day operational tasks, middle managers provide the bridge between them through using the vision and ideas created by top management, converting them into more concrete concepts, and then helping front-line employees understand and implement them effectively. At the same time, when front-line employees encounter a problem or challenge in their tasks, they create new knowledge which can then be converted by middle-managers into broader concepts and ideas for the organisation as a whole, and again passed on to the top to feed back into the wider vision and strategy. New knowledge therefore constantly moves up and down through the high involvement of middle managers in both directions. Table 5.1 shows a summary of the three types of organisations, in terms of *who* is involved in managing knowledge, the structure behind it, *what* kind of knowledge is created and managed, and an event organisation example for each type.

Table 5.1: Comparison of top-down, bottom-up and middle-up-down approaches to knowledge management

	Top-down	Bottom-up	Middle-up-down
Knowledge creation through...	Top management	Entrepreneurial individuals	Teams and middle managers (knowledge engineers)
Structure	Hierarchical	Project teams and/or informal networks	Combination of hierarchy and task force
Focus on	Explicit knowledge stored in databases, manuals or checklists	Tacit knowledge embodied by individuals	Combination of explicit and tacit knowledge within an organisational knowledge base
Event organisation example	Directors, managers, or permanent staff in charge of decision-making	Seasonal staff and volunteers largely autonomous	Team or project leaders as coordinators

In order for the middle managers to manage knowledge up and down the organisation, a typical hierarchical structure would be ineffective. Similarly, these fixed structures pose a limitation on the relationships required for effective knowledge practices to develop. Furthermore, in many organisations, large amounts of knowledge do not necessarily sit within a certain department or functional team, but rather within communities of people or communities-of-practice. They tend to have some overlapping knowledge in common, but they also possess large amounts of specialised and task-specific knowledge (Carlile, 2002, 2004). One way of creating slightly more formal communities is through interdisciplinary or cross-functional teams that work together on specific tasks.

Interdisciplinary teams and pods

The structure of an organisation is generally influenced by its culture and emphasis on collaboration and innovation (Chen & Huang, 2007). A decentralised and more integrated organisational structure, for example, has a positive impact upon the organisational culture, communication, social interactions and therefore relational knowledge management (Gorelick et al., 2004; McLean, 2005; Chen & Huang, 2007). In the broader literature, Albers and Brewer (2003) and Fenton and Albers (2007) further highlighted the importance of group structures that focus on diversity among group members to enhance knowledge creation and transfer. Formal as well as informal groups, or pods, within an organisation are examples of such structures that can enhance relational knowledge management (Wenger, 1998; Cabrera & Cabrera, 2005; Fenton & Albers, 2007). Through the creation of these formal and informal groups, knowledge flows not only vertically, but also horizontally.

Diversity in terms of backgrounds and knowledge within an interdisciplinary team can increase the team's generation of new ideas and thus enhance creativity and innovation (McLean, 2005; Chen & Huang, 2007). Whereas a shared vision and understanding provides the basis for effective knowledge transfer, Ferran-Urdaneta (1999: 131) argued that, "the more homogenous the teams is [sic], the less effective it will be for knowledge creation." To increase innovation and the transfer of know-how across functional areas organisations can, for example, create cross-functional teams composed of a diverse group of people with different areas of expertise (Currie & Kerrin, 2003). In event and festival organisations, seasonal staff members are usually put together in teams around functional areas, such as technical staff forming a team and producers forming another team (Van der Wagen, 2007). An emphasis on functional teams, however, makes it difficult to effectively transfer know-how across the teams, particularly when the tasks are complex. Here, teams of people with diverse backgrounds can be more beneficial in terms of sharing knowledge effectively and efficiently than homogenous teams. Oborn and Dawson (2010: 1836) argued that, "[c]onjoining expertise between colleagues from different backgrounds can enable novel ways of distinguishing and connecting ideas." They can either informally emerge as a community-of-practice as discussed in Chapter 4, or they can be formally structured, such as in interdisciplinary pods.

The case study below presents one such example of interdisciplinary pods within an Australian music festival organisation.

Case Study: Pod structure at the Queensland Music Festival, Australia

The pod structure discussed in this chapter can be applied to the Queensland Music Festival (see earlier introduction to the case study in Chapter 4): within the 2011 QMF team, the permanent staff each held one major functional responsibility, including executive director, artistic director, finance and administration manager, program director, technical director, and marketing and development director. The seasonal staff were then set up in several pods, each consisting of a producer, a project coordinator and a technical manager as well as one or two interns during the festival season. Each pod was responsible for a number of events with their own network of contractors, creatives and artists. Furthermore, there was a marketing professional associated with each event, thus the different pods worked together with a centralised marketing team as well (see Figure 5.2). The interdisciplinary pod structure is unique to QMF and quite different from the more traditional structure around functional areas usually found in festival and event organisations. However, throughout the knowledge transfer practices and processes, pod members were able to translate different knowledges (artistic, technical, administrative, operational, strategic) and came to understand how their know-how contributed to the team. A member of one of the pods explained how they worked together as a team and how important it was to have teams of people with different areas of expertise (interview 07 June 2011; names have been changed):

> So the three people working together, me, Veronica and Claire ... there's a lot of experience put together. Which goes to show how this organisation has done its set-up in the pods ... the [other] festivals I've been to and worked with don't do that. They seem to clump technical together, they seem to clump producers together. Now, that makes absolutely no sense. If you drew that on a diagram, it makes no sense, because ... why? As a technical [manager], I don't need to talk to my other technical managers. I need to talk to my direct show! Our four shows, we talk together. If I need to get information from other tech-

nical managers, I stand up, walk over and talk to them. But more than likely, I will be talking to the other two people on my show: the producer and the project coordinator (...) So, it's a very good set-up in that way and not many people do that, which kind of shocks the hell out of me.

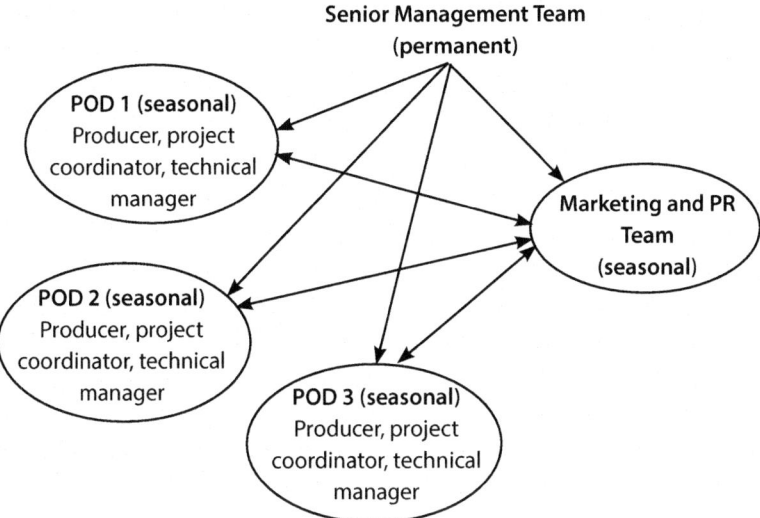

Figure 5.2: Interdisciplinary pod structure at QMF

Within the pods staff members developed necessary know-how around their individual projects, while at the same time they could also consult other professionals outside their pod if necessary. In project-based organisations this interdisciplinary structure is quite common and a very effective way of sharing expert knowledge within, as well as across, the teams and the organisation as a whole. It is commonly associated with a decrease in knowledge hoarding in functional silos. The QMF office was furthermore designed as an open office with only a few private offices and meeting rooms. Most staff members were located in the main room and sat together in these teams and pods, so they could communicate easily. The QMF spatial office layout and interdisciplinary pod structure therefore contributed to informal exchanges of information and knowledge, and helped staff members engage in a range of knowledge practices both within as well as across the teams and pods.

(Based on research by Stadler, 2013)

Knowledge management roles and responsibilities

Within the wider knowledge management literature, several knowledge management roles and responsibilities have been identified at different levels within the organisation. In some organisations these are explicitly defined for employees as part of their job, however, this is not usually the case in event organisations. Here, some employees and staff members perform the different roles, but this is an implicit part of their job and not usually stated in their job descriptions (Stadler et al., 2014). In other words, event staff members would not usually refer to themselves as, for example, 'knowledge brokers' (see below).

Knowledge management roles can be performed at different levels and therefore come with different responsibilities at each level. Some of the terms used in the knowledge management literature are confusing or used interchangeably, but the three most commonly identified roles are; chief knowledge officers, knowledge brokers and knowledge workers (Earl & Scott, 1999; Meyer, 2010):

- *Knowledge officers* (Nonaka & Takeuchi, 1995) or chief knowledge officers (Earl & Scott, 1999; Bergeron, 2003; Schuett, 2003) are usually top managers and responsible for managing the entire organisational knowledge management strategy and processes at the corporate level. They identify strategies, design systems and facilitate the transfer of explicit as well as tacit knowledge.

- Similar to chief knowledge officers, *knowledge management champions* and strategists (Burstein et al., 2010) are part of the senior management team and therefore responsible for the overall vision of the company. However, while there usually only is one chief knowledge officer in an organisation, there can be more than one knowledge management champion and strategist.

- *Knowledge engineers* (Nonaka & Takeuchi, 1995) or *knowledge brokers* (Meyer, 2010) tend to be middle managers and they serve as a bridge between the visionary ideals of the top and the day-to-day operational tasks of front-line workers. In terms of knowledge management, knowledge brokers are therefore responsible for creating connections between different people (and between different levels) of the organisation so that they can share knowledge, as well as actually move knowledge around between them. They know the networks within the organisation well and are also aware of the networks and links with other (internal and external) stakeholders. Their role in creating relational knowledge practices is therefore absolutely crucial.

- Brown and Duguid (1998) further distinguished between *knowledge brokers* and *knowledge translators,* whereby brokers are members of different, but overlapping, communities (or communities-of-practice) and use their knowledge of these two or more communities to facilitate mutual understanding between them. In other words, a knowledge broker is able to transfer knowledge of a particular practice from one community to another. Knowledge translators, on the other hand, work across mutually exclusive communities and 'translate' or reframe the ideas of one community to then make them accessible and understandable to another.

- *Knowledge practitioners* (Nonaka & Takeuchi, 1995) or *knowledge workers* (O'Dell, 2004; Debowski, 2006; Burstein et al., 2010) are front-line workers, who generally speaking, "use their heads more than their hands to produce value" (Horibe, 1999: xi). The roles and tasks of knowledge workers are primarily intellectual, creative, and/or non-routine; these tasks require problem-solving skills and the creation and use of knowledge (Hislop et al., 2018). In any knowledge-intensive firm or organisation therefore, everybody can be considered a knowledge worker. Knowledge workers create, share and use knowledge, and they engage in a range of knowledge practices as part of their job. They embody, accumulate, generate and update both tacit and explicit knowledge.

- Nonaka and Takeuchi (1995) further distinguished between *knowledge operators* here, who mainly deal with tacit knowledge; and *knowledge specialists*, who mainly deal with explicit knowledge.

These knowledge management roles have been applied to a festival organisation by Stadler et al. (2014), who identified a number of knowledge champions and strategists, knowledge brokers and knowledge workers within their case organisation. They argued that a more explicit understanding of these roles and responsibilities can positively contribute to the professionalisation of the events industry, in the sense that, if staff members are aware of their knowledge management roles and what they entail, they can more effectively and more efficiently identify those knowledge practices that enhance the creation, transfer and utilisation of knowledge, and at the same time negotiate and reshape the ones that do not contribute to the overall success of the organisation.

In addition to the roles identified above, leadership and knowledge management also go hand-in-hand. It would be beyond the scope of this book to fully cover leadership, but it is commonly argued that "leaders in learning organisations are required to be learners as much as teachers, and they should also have roles as coaches and mentors. Such a leadership style is necessary not only to actively stimulate the curiosity and learning of new workers, but to also make leaders sensitive and responsive to the opinion of workers" (Hislop et al., 2018: 100). While strong leaders in an organisation are important for creating a good culture and shared values (see Chapter 6), they can also take on specific roles in relation to knowledge management, particularly in small companies and SMEs, such as event organisations (Clayton, 2020). It is commonly argued that effective leadership is crucial for organisational learning as well as to enhance knowledge activities and knowledge practices within an organisation. A lot of this is down to 'leading-by-example' and therefore leaders can be regarded as role models in terms of what kind of knowledge to share, how to share it, how to work with other people in the organisation, and how to create effective communication and collaboration. Leaders are also important with regards to inspiring and motivating other employees, rather than micro-managing them (Hislop et al., 2018). They can, for example, help interpret, nurture and support knowledge practices throughout the organisation, as well as help create new knowledge rather than merely control and direct it. Debowski (2006: 62) argued that, "there are some essential contributions knowledge leaders make that rely on an understanding of systems, processes and people, and shape them into a cohesive and functional whole. They include the capacity to provide strategic visions, motivate others, effectively communicate, act as a change agent, coach others, model good practices, and carry out the knowledge agenda." She (2006: 68) identified three key contributors and their roles as, (1) *strategic knowledge leaders* (providing vision and direction for the knowledge management strategy); (2) *core leaders* (managing and controlling the outputs and focus of organisational units); and (3) *team leaders* (supporting and encouraging teams and their activities).

Finally, it is important to acknowledge that in event organisations, similar to other knowledge-intensive organisations, some knowledge-intensive tasks might need to be outsourced and therefore at times, other stakeholders need to take on specific knowledge management roles and responsibilities as well. For example, event organisations do not generally possess all the knowledge of how to set up a stage, or the lighting and sound systems required for a performance. These tasks require specialist knowledge from a production company and can therefore be outsourced rather than carried out in-house. With the large number of suppliers and contractors necessary to run a successful event, outsourcing knowledge-intensive tasks is a common part of effective knowledge management in event organisations. Stakeholder theory helps with identifying those internal and external stakeholders who are key to the process (see van Niekerk & Getz, 2019), and their specific knowledge management roles and responsibilities can be developed from there. On an even more complex level, when cities and events create event portfolios, the stakeholder network becomes even larger and includes inter-organisational relationships, as well as strategic collaborations (Antchak, Ziakas, & Getz, 2019). These could also benefit from making knowledge management roles and responsibilities within the network more explicit and hence enhancing the knowledge transfer process between the different actors and institutions. The roles of knowledge brokers are crucial in this.

Study and discussion questions

- ☐ Explain and discuss the difference between top-down and bottom-up knowledge management approaches. Debate which one is better suited for event organisations and why?
- ☐ What kind of knowledge in an event organisation can be 'outsourced' to external stakeholders? Explore the challenges of 'outsourcing' knowledge using event-related examples.
- ☐ A shared office space can enhance the transfer of knowledge within and between work teams. Are there other ways of creating 'shared spaces' in an event organisation? Who should have access to these 'spaces' and why?
- ☐ Can all employees in an event organisation be considered 'knowledge workers'? What about volunteers? Discuss and provide examples from your own experience.

Recommended additional readings

Cabrera, E. F., & Cabrera, A. (2005). Fostering knowledge sharing through people management practices. *The International Journal of Human Resource Management*, 16(5), 720-735.

Lockstone-Binney, L., Hanlon, C., & Jago, L. (2020). Staffing for successful events: Having the right skills in the right place at the right time. In S. J. Page & J. Connell (Eds.), *The Routledge Handbook of Events* (2nd ed., pp. 427-441). London: Routledge.

Stadler, R., Fullagar, S., & Reid, S. (2014). The professionalization of festival organizations: A relational approach to knowledge management. *Event Management*, 18(1), 39-52.

References

Albers, J. A., & Brewer, S. (2003). Knowledge management and the innovation process: The eco-innovation model. *Journal of Knowledge Management Practice*, 4(6), 1-6.

Allen, J., O'Toole, W., McDonnell, I., & Harris, R. (2011). *Festival and Special Event Management* (5th ed.). Milton, Qld.: John Wiley & Sons.

Antchak, V., Ziakas, V. & Getz, D. (2019). *Event Portfolio Management*. Oxford: Goodfellow Publishers.

Arcodia, C. (2009). Event management employment in Australia: a nationwide investigation of labour trends in Australian event management. In T. Baum, M. Deery, C. Hanlon, L. Lockstone, & K. Smith (Eds.), *People and Work in Events and Conventions - A Research Perspective* (pp. 17-28). London: CABI.

Bergeron, B. (2003). *Essentials of Knowledge Management*. Hoboken NJ: John Wiley & Sons, Inc.

Brown, J. S., & Duguid, P. (1991). Organizational learning and communities-of-practice: Toward a unified view of working, learning, and innovation. *Organization Science*, 2(1), 40-57.

Burstein, F., Sohal, S., Zyngier, S., & Sohal, A. S. (2010). Understanding of knowledge management roles and responsibilities: a study in the Australian context. *Knowledge Management Research & Practice*, 8(1), 76-88.

Cabrera, E. F., & Cabrera, A. (2005). Fostering knowledge sharing through people management practices. *International Journal of Human Resource Management*, 16(5), 720-735.

Carlile, P. R. (2002). A pragmatic view of knowledge and boundaries: boundary objects in new product development. *Organization Science*, 13(4), 442–455.

Carlile, P. R. (2004). Transferring, translating, and transforming: an integrative framework for managing knowledge across boundaries. *Organization Science*, 15(5), 555-568.

Chen, C.-J., & Huang, J.-W. (2007). How organizational climate and structure affect knowledge management - The social interaction perspective. *International Journal of Information Management,* 27(2), 104-118.

Clayton, D. (2020). Knowledge management in events. In S. J. Page & J. Connell (Eds.), *The Routledge Handbook of Events* (2nd ed., pp. 442-456). London: Routledge.

Cranwell-Ward, J., & Abbey, A. (2005). *Organizational Stress.* New York: Palgrave Macmillan.

Currie, G., & Kerrin, M. (2003). Human resource management and knowledge management: enhancing knowledge sharing in a pharmaceutical company. *International Journal of Human Resource Management,* 14(6), 1027-1045.

Debowski, S. (2006). *Knowledge Management.* John Wiley & Sons Ltd.

Du Plessis, M. (2006). *The Impact of Organisational Culture on Knowledge Management.* Oxford: Chandos Publishing.

Earl, M. J., & Scott, I. A. (1999). Opinion: What is a chief knowledge officer? *Sloan Management Review,* 40(2), 29-38.

Fenton, D., & Albers, J. A. (2007). Leveraging knowledge in the sales force of a pharmaceutical company. *Journal of Knowledge Management Practice,* 8(4).

Ferran-Urdaneta, C. (1999). *Teams or Communities? Organizational Structures for Knowledge Management.* Boston University Systems Research Center.

Getz, D. (2018). *Event Evaluation.* Oxford: Goodfellow Publishers.

Gloet, M., & Berrell, M. (2003). The dual paradigm nature of knowledge management: implications for achieving quality outcomes in human resource management. *Journal of Knowledge Management,* 7(1), 78-89.

Gorelick, C., Milton, N., & April, K. (2004). *Performance through Learning - Knowledge Management in Practice.* Amsterdam: Elsevier Butterworth-Heinemann.

Hanlon, C., & Cuskelly, G. (2002). Pulsating major sport event organizations: a framework for inducting managerial personnel. *Event Management,* 7(4), 231-243.

Hislop, D., Bosua, R., & Helms, R. (2018). *Knowledge Management in Organizations - A Critical Introduction* (4th ed.). Oxford: Oxford University Press.

Horibe, F. (1999). *Managing knowledge workers: New skills and attitudes to unlock the intellectual capital in your organization*: John Wiley & Sons.

Küpers, W. (2005). Phenomenology of embodied implicit and narrative knowing. *Journal of Knowledge Management,* 9(6), 114-133.

Lockstone-Binney, L., Hanlon, C., & Jago, L. (2020). Staffing for successful events: Having the right skills in the right place at the right time. In S. J. Page & J. Connell (Eds.), *The Routledge Handbook of Events* (2nd ed., pp. 427-441). London: Routledge.

McLean, L. D. (2005). Organizational culture's influence on creativity and innovation: a review of the literature and implications for human resource development. *Advances in Developing Human Resources*, 7(2), 226-246.

Meyer, M. (2010). The rise of the knowledge broker. *Science Communication*, 32(1), 118-127.

Mosley, D. C., Megginson, L. C., & Pietri, P. H. (2001). *Supervisory Management - The Art of Empowering and Developing People* (5th ed.). Mason OH : Thomson South-Western.

Nankervis, A., Compton, R., Baird, M., & Coffey, J. (2011). *Human Resource Management - Strategy and Practice* (7th ed.) CENGAGE Learning Australia.

Nonaka, I., & Takeuchi, H. (1995). *The Knowledge Creating Company – How Japanese Companies Create the Dynamics of Innovation*. New York: Oxford University Press.

Oborn, E., & Dawson, S. (2010). Knowledge and practice in multidisciplinary teams: Struggle, accommodation and priviledge. *Human Relations*, 63(12), 1835-1857.

O'Dell, C. (2004). *The Executive's Role in Knowledge Management.* Houston: APQC Publications.

Robertson, M., & Swan, J. (2004). Going public: The emergence and effects of soft bureaucracy within a knowledge-intensive firm. *Organization*, 11(1), 123-148.

Schuett, P. (2003). The post-Nonaka knowledge management. *Journal of Universal Computer Science*, 9(6), 451-462.

Stadler, R. (2013). Power relations and the production of new knowledge within a Queensland Music Festival community cultural development project. *Annals of Leisure Research*, 16(1), 87-102.

Stadler, R., Fullagar, S., & Reid, S. (2014). The professionalization of festival organizations: A relational approach to knowledge management. *Event Management*, 18(1), 39-52.

Van der Wagen, L. (2007). *Human Resource Management for Events – Managing the Event Workforce.* Amsterdam: Elsevier.

Van der Wagen, L., & White, L. (2014). *Human Resource Management for the Event Industry* (2nd ed.). London: Routledge.

Van Niekerk, M., & Getz, D. (2019). *Event Stakeholders*. Oxford: Goodfellow Publishers.

Wenger, E. (1998). *Communities of Practice - Learning, Meaning, and Identity.* Cambridge: Cambridge University Press.

Werner, J. M., & DeSimone, R. L. (2006). *Human Resource Development* (4th ed.). Mason OH : Thomson South-Western.

Yahya, S. & Goh, W.-K. (2002). Managing human resources toward achieving knowledge management. *Journal of Knowledge Management*, 6(5), 457-468.

6 Cultural Elements of Knowledge Management

Learning objectives

- ☐ Define organisational identity, organisational vision and organisational culture.
- ☐ Understand the importance of creating an open, collaborative culture to support knowledge practices.
- ☐ Explore motivation and trust within event organisations as crucial factors underpinning the knowledge management process.
- ☐ Discuss the importance of collaboration and co-creation in work tasks as well as for effective knowledge practices.

Introduction

Organisational culture is, perhaps not surprisingly, by far the most researched topic in relation to knowledge management to date. It is widely argued that an open, collaborative culture enhances knowledge processes, activities and practices, and that this open culture will help organisations be successful in the long-run (see for example, Du Plessis, 2006; Kathiravelu et al., 2014; Intezari et al., 2017). Organisational values, assumptions, and the cultural context shape what employees believe in, their shared understanding of how things are done in the organisation, as well as their shared language. The process of meaning-making through different knowledge practices is therefore largely shaped by organisational culture and embedded in it (Hislop et al., 2018).

In the wider knowledge management literature, there are five characteristics of an organisational culture that have been identified as contributing to and facilitating knowledge activities and practices (Donate & Guadamillas, 2011; O'Dell & Hubert, 2011):

- Knowledge sharing is regarded as the norm within the organisation;
- Staff members have a strong sense of organisational identity;
- There is a high level of trust and respect between staff members;
- Organisational processes are regarded as fair and transparent; and
- Staff members have a high level of trust in management.

Based on these assumptions, this chapter builds on some of the structural elements of knowledge management highlighted in Chapter 5, but provides a more in-depth discussion of the cultural context within which they are performed and practised. For example, when recruiting and selecting employees, it is important to take their background, values and assumptions into account and to see whether these match the organisation's vision and values. If this is the case, then new employees will be more likely to positively contribute to the organisation's culture and identity, and they will be able to more effectively practise knowledge management (Hislop et al., 2018). The chapter starts with a discussion of organisational identity and critically explores how employees identify with an organisation. This will naturally lead into a section on organisational culture, where beliefs, values, and behaviours shared by all employees in an organisation can also positively contribute to their ways of working, their shared understanding, and shared processes of meaning-making. The second part of the chapter then more specifically highlights motivation and trust in event organisations, as well as collaboration and co-creation when engaging in knowledge practices, which all tend to be higher in open organisational cultures and therefore positively contribute to knowledge management (Du Plessis, 2006).

Organisational identity

Ashforth et al. (2008) maintained that identity is relational and is constituted through organisational members' ways of identifying with the organisation. By stating that 'I am a member of the organisation and it is important to me', people feel positively about their membership and are emotionally invested in being part of the organisation. Identification with an organisation therefore helps employees articulate their values, goals, beliefs, as well as knowledge, skills and abilities that shape their behaviours and practices within the organisation (Ashforth et al., 2008). The concept of organisational identity is further useful for understanding knowledge practices that are central to *why* people join a certain organisation, *how* they approach their work and how and why they interact with each other when performing their work and tasks. It includes the way they *feel* about the organisation as a whole, and can therefore be quite a fuzzy concept to explain to others or to newcomers to the organisation.

In terms of knowledge management, the extent to which an employee identifies with the organisation as a whole, or at least with their team, pod, or community-of-practice, usually positively contributes to their willingness to share knowledge with others (Hislop et al., 2018). A strong organisational identity also helps create and enhance social relationships within the organisation and therefore the creation of mutual trust, which is crucial for effectively managing knowledge, as discussed further below. Addleson (2012: 8) argued that, "[p]eople 'share knowledge' best when they work at creating a context of shared meanings for one another in their conversations." Hence, a shared understanding of identity ('who we are') provides the basis for effective collaboration. In event organisations, on the one hand, it can be difficult to create and maintain a shared identity due to their temporary, 'pulsating' nature. Without such a shared event identity there is no common ground for knowledge to be effectively practised. On the other hand, event organisations sometimes do have a history beyond each event season and therefore an established identity that exists over time. Creating a strong organisational identity over time can develop a sense of organisational loyalty among employees, which is crucial in events in terms of retaining permanent staff, and getting seasonal staff members and volunteers to come back each event season. A certain sense of pride in the event and hence loyalty to it thereby helps in retaining these employees.

There are three characteristics commonly identified that together construct an organisational identity (Deal & Kennedy, 1982; Albert & Whetten, 2004) within which organisational knowledge can be practised and shared; (1) the essence of the organisation; (2) what distinguishes the organisation from other organisations; and (3) continuity over time. Some of this can of course be expressed through an organisation's vision or mission statement. An organisational vision describes a shared meaning (Kelly, 2000) of what the organisation is, as well as "the organization's central characteristics as a guide for what they should do and how other institutions should relate to them" (Albert & Whetten, 2004: 92). Furthermore, the vision provides the foundation for a shared understanding among the event staff and board members, sponsors, partners, councils, artists and community members. It is therefore important to share this vision not just internally, but also with external stakeholders and partners in order to create a shared language (Orlikowski, 2002; Renzl, 2007; Hecker, 2012) that reinforces the importance of what the event organisation does and what it stands for, as well as a shared context for knowledge to be created and transferred. Organisational members can then learn 'how to' perform their roles within the organisation and 'how to' effectively work together and collaborate.

A strong organisational identity, vision and mission are therefore key contributing factors to effective knowledge management. However, too strong a sense of organisational identity can potentially lead to exclusivity, group think and cliquey-ness in organisations, where it then becomes increasingly difficult for new employees to join the team and learn how to contribute to the organisation.

Organisational culture

While the organisational identity defines 'who we are' as an organisation, organisational culture describes the 'way things are done' within the organisation. It can be defined as "the beliefs and behaviours shared by an organization's members regarding what constitutes an appropriate way to think and act at work" (Hislop et al., 2018: 273), and includes ideas and values, as well as certain acceptable behaviours (Alvesson & Sveningsson, 2008). Furthermore, it needs to be shared by a significant number of people; ideally all members of the organisation.

Elements or categories of organisational culture have been discussed by many authors and in many different ways. It would be beyond the scope of this book to discuss them all in detail, but Figure 6.1 is based on Schein's (2004) work, and will be applied to events and explained using an event-related example. In some organisations one or two of these elements of organisational culture are more important than others, while in other organisations there is a good mix of all of the below. The case study on page 120 provides some ideas of which cultural elements work well for a festival organisation.

Observed behavioural regularities when people interact	For example, the language they use, or the customs and traditions that evolve
Group norms	Tacit and often taken-for-granted standards and values that evolve in working groups and teams
Espoused values	The articulated and announced principles and values of an organisation that are shared with external partners and the wider public
Formal philosophy	Policies, principles and ideologies that guide an organisation's actions towards its stakeholders
Rules of the game	The tacit and unwritten 'way things are done' within the organisation. These usually need to be learned by all new employees to become accepted members of the organisation
Climate	The way employees interact with each other and with other stakeholders, as well as the physical layout of the workplace (e.g. open offices, shared offices)
Embedded skills	Special knowledge and competencies that area passed on from one generation to the next, but are not necessarily written down
Habits of thinking, mental models, and linguistic paradigms	Cognitive frames and language used by the members of the organisation, which are usually taught to newcomers early on
Shared meanings	Understandings within a group or team, or even the organisation as a whole, that evolve when employees regularly interact with each other
'Root metaphors' or integrating symbols	Ways of working together that reflect the emotional and aesthetic values of the organisation, sometimes embodied in artefacts, buildings, or offices
Formal rituals and celebrations	Organisational ways of celebrating key events and milestones, such as promotions, but also informal rituals such as a staff member's birthday

Figure 6.1: Categories of organisational culture

It can be seen that most, if not all, of these categories describe tacit ways of doing things or tacit ways of working with others, and therefore constitute some of the 'soft' factors for knowledge management. They cannot be learned from books, manuals or induction kits, but rather need to be absorbed and acquired over time through stories, narratives, learning-by-doing, or observing others in the organisation. In that sense, they are key practices that form part of the tacit knowledge that is necessary for employees to create, share and use in their work. What works for one organisation might not necessarily work for another. For example, while some common language exists in event management around the logistical and operational tasks necessary to put together a successful event, the language used to convey these, the way employees express their importance and priority over other tasks, and the way other stakeholders are engaged in the process will be different from one organisation to another. A basic understanding can be taught, for example, at university, in training and workshop sessions, or inductions, but a lot of this tacit understanding of how things are done in one event organisation versus another needs to be learned over time through informal knowledge processes and practices. The fact that many event staff members move on to other organisations once the event is over, makes this more difficult than in other, more permanent organisations, as they will need to learn and relearn different languages and different ways of doing things each time.

Du Plessis (2006: 61) further emphasised that, "[k]nowledge management should be implemented as an embedded part of the knowledge management and organisational culture of the organisation by making it an integral part of the way people work on a daily basis and not by making it something 'extra' that they have to do over and above their normal workload." In other words, in order for knowledge management strategies to be successfully implemented in an organisation, it is beneficial to align them with the existing organisational culture. For example, a new knowledge management initiative should always be focussed on existing organisational problems and matched with existing 'ways of doing things' within the organisation. Reward and appraisal systems should also be in place, for example to show what kind of knowledge creation and sharing behaviour is encouraged within the organisation, and what should be avoided. Lastly, knowledge management initiatives should also link into existing networks of relations within the organisation (Hislop et al., 2018).

Debowski (2006) further discussed different knowledge culture enablers and summarised them along the lines of, (1) core values, such as collaboration, communication, innovation, trust, and a mutual understanding that knowledge is valued and should be shared; (2) structural support (the structure of the organisation itself, e.g. hierarchical or non-hierarchical, human resource management, problem solving and transparent decision-making, access to information and communication channels); (3) enacted values, including leaders, models, as well as opportunities and encouragement to collaborate; and finally (4) interaction with colleagues (mentorship, team behaviour, and the quality and focus of co-worker interaction).

Similarly, in a more recent review of the knowledge management literature, Intezari et al. (2017) identified seven cultural factors that affect knowledge processes, including knowledge creation, sharing, and implementation in many different ways. These are specified as: social interactions; openness in communication; trust; perception of knowledge; top management's support and involvement; freedom vs control; and compensation system (organisational rewards). For a full discussion of the factors and their relation to knowledge management processes, see Intezari et al. (2017: 503-506). Interestingly, the authors note that these cultural factors are mainly linked to the process of knowledge sharing, which is perhaps in line with the wider trend in knowledge management research, where sharing is by far the most researched topic, as discussed in earlier chapters. Intezari et al. (2017) call for a more integrated approach to investigating knowledge cultures that includes all knowledge management activities, processes and practices.

Within the event management literature, organisational culture has been discussed in a number of academic papers on events and knowledge management, which have previously been referred to in this book (for further reading on this topic, see for example, Abfalter et al., 2012; Stadler et al., 2013; Ragsdell & Jepson, 2014; Stadler et al., 2014; Clayton, 2016; Stadler & Fullagar, 2016; Muskat & Deery, 2017); as well as a source for creativity and innovation (e.g., Carlsen et al., 2010; Larson, 2011); and is generally mentioned as a positive element for the success of an event in Events Management textbooks more broadly. Getz (2002: 216) specifically highlights 'organisational culture' (or lack thereof) as a potential factor of why festivals fail, and stresses that learning organisations with clear strategic planning and research functions are "more likely to be adaptable" than temporary event organisations.

Lastly, it is worth mentioning that in event organisations, the 'family' metaphor is often used not only in describing a group of attendees or fans, but also for the staff and volunteers who share working together on the event experience (Smith & Lockstone, 2009; Van der Wagen & White, 2014); however, the importance of belonging to a 'family' has not yet been identified as a knowledge enabler within event organisations. The term 'family' is thereby used to describe a certain organisational culture that features a family ideology (Ram & Holliday, 1993). This 'family' metaphor is however not used to describe a normative family, but rather a family by choice. It not only describes a sense of belonging to the organisation, but also terms such as 'trust', 'caring', 'responsibility to each other', 'continuation of community' and 'commitment' (Lennon & Wollin, 2001: 418). It has been argued that these feelings and emotions, in turn, positively influence a collaborative culture, knowledge management and in particular knowledge sharing practices (Hislop, 2003; Donate & Guadamillas, 2011).

Case Study: The Queensland Music Festival vision, identity and culture

The Queensland Music Festival was introduced in the case study in Chapter 4. Its vision is:

To transform lives through unforgettable musical experiences

Below are a few interview quotes and snippets from the festival promotional material to show how this vision was being communicated and how staff members felt about it:

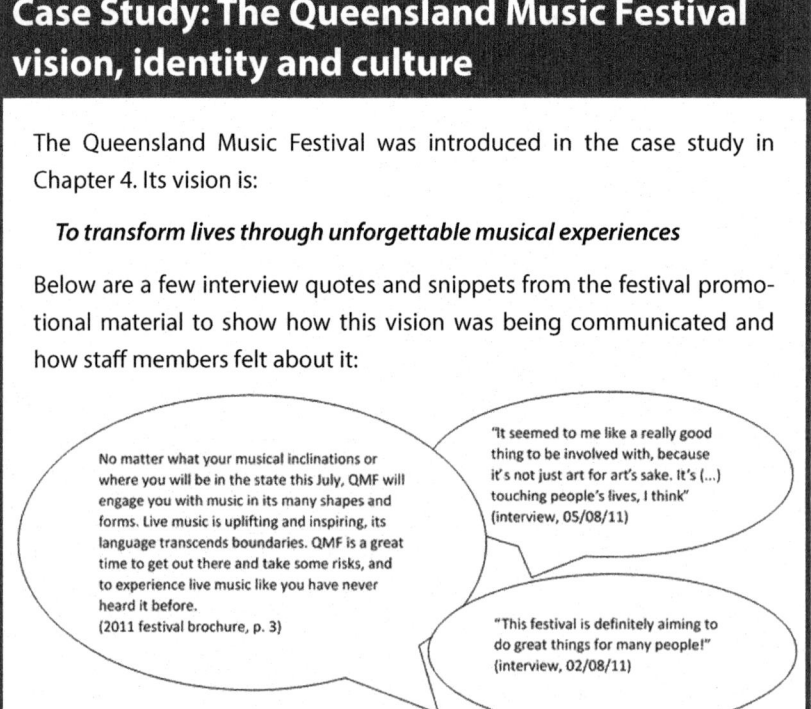

No matter what your musical inclinations or where you will be in the state this July, QMF will engage you with music in its many shapes and forms. Live music is uplifting and inspiring, its language transcends boundaries. QMF is a great time to get out there and take some risks, and to experience live music like you have never heard it before.
(2011 festival brochure, p. 3)

"It seemed to me like a really good thing to be involved with, because it's not just art for art's sake. It's (...) touching people's lives, I think"
(interview, 05/08/11)

"This festival is definitely aiming to do great things for many people!"
(interview, 02/08/11)

6: Cultural Elements of Knowledge Management

As discussed throughout this chapter, however, creating a vision for an organisation alone is not enough. Event staff members (both permanent and seasonal) also need to buy into this vision in order to learn how to contribute to it in their various roles. At QMF, the senior management team were aware of this and very proud of how staff members have embraced the festival vision; while some of the seasonal staff and even volunteers seemed inspired by what the festival aims to do and see this as the main reason why so many people in the industry want to work for QMF. As mentioned in the chapter above, this can then lead to a shared organisational identity and loyalty:

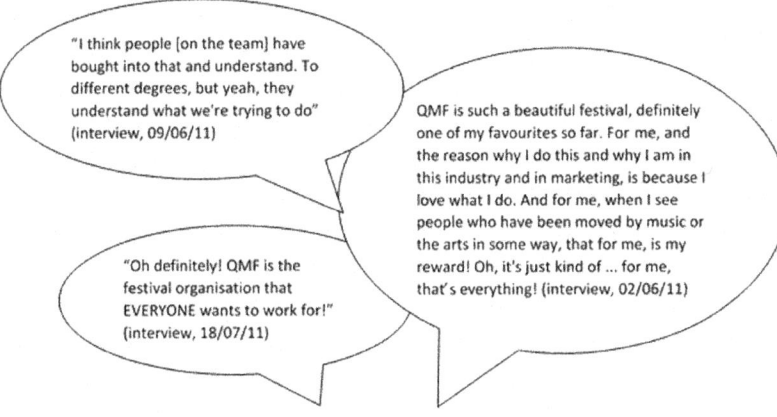

Lastly, based on the shared vision and identity, which staff members very much identified with in this festival organisation, they created an open and collaborative culture in which all festival members felt comfortable to contribute new ideas. This nurturing context was crucial for the development of effective work relationships, trust and mutual respect, which are all elements of a knowledge-friendly culture. Therefore, at QMF innovation was not merely the artistic and executive directors' responsibility, but rather achieved through sharing knowledge among the entire team and creating new knowledge together. Sharing knowledge was not a one-directional or bottom-up process either, but rather the senior and permanent team shared everything with everyone and encouraged all members of the organisation to share their ideas with them in turn. There were no secrets. Authority was thus distributed among the entire team, not used by the core team to impose their ideas upon seasonal staff members and volunteers. The openness embodied by the permanent staff created a friendly and supportive atmosphere and culture where seasonal staff members understood the importance of collaboration and willingly shared their knowledge with each other as well as with the permanent staff:

> I think it is a very good atmosphere, also a very empowering atmosphere in the core team, which means that (...) you can say what you think and you can possibly influence things in a way which means that you have a lot of great minds thinking alike and you get a much better outcome. As opposed to just [them] saying "this is what you've got to do" (interview, 05/08/11).

> Oh our conversations in the marketing room... We just yell out! (laughs) There are no secrets here! You know, I wheel back and go, "hey... what do you think about this? Let's have a chat..." And we all start talking, it's great. (...) The more people get involved, I mean someone else might have a better idea than you. So bring it on, let's all talk about it! (interview, 02/06/11)

> I think, what's good about the team process here is that people share ideas and knowledge and out of that process, you get these little nuggets of gold that turn into something like an entirely new project! (interview, 15/06/11)

For many of the staff members, whilst working in the highly stressful and intense festival environment, they started spending more time with their work colleagues than with their families and friends. This helped develop close relationships among the team, where some even started saying, "It's like a family away from my own family." The family metaphor frequently used by QMF staff members nicely summarises the way they worked together, as expressed by this staff member:

> "Oh yes, we are a bit like a big family. We spend so much time together, we care and look out for each other, but we also sometimes bicker and argue like you would argue with your brothers or sisters, I suppose... [laughs] I think it just shows how close we all are and the relationships we've built across the team. I'm proud to be a member of this family, they're good people; hard-working, passionate, always happy to help!" (interview, 22/06/11)

(Based on research by Stadler, 2013)

Motivation

"Whatever approach to knowledge management an organization adopts, the motivation of workers to participate in such processes will be key to their success" (Hislop et al., 2018: 177). If employees identify with the organisation, and are highly committed to it, they will be more motivated to share knowledge with others. Organisational commitment means that employees will feel some sort of emotional attachment to the organisation, and more positively align their values and goals with those of the organisation. Motivation is therefore a key element of knowledge management that can help tackle some of the challenges mentioned in Chapter 2. Employees will be more motivated to engage in knowledge practices, if they feel that positive benefits will outweigh negative ones; while at the same time, if the negative outcomes are perceived to be greater than the positive ones, employees will probably start hiding or hoarding their knowledge and not participate in knowledge practices (Hislop et al., 2018).

The much discussed distinction between intrinsic and extrinsic motivation also applies to employees' motivation to create, share, use and store knowledge effectively, where intrinsic motivation is usually mentioned as more important than extrinsic motivation (Cruz et al., 2009). People are extrinsically motivated when they expect a tangible benefit or for example, a certain amount of money for their achievements. For paid staff in event organisations, these can be bonuses, promotions, or free tickets. But also for volunteers there are many extrinsic motivations, such as free admission to the event, free merchandise, certificates, acknowledgements, social occasions and parties, or meeting guest artists and performers (Allen et al., 2011; Muskat & Deery, 2017). Intrinsic motivation, on the other hand, occurs when people are motivated out of a personal goal; in other words, the content of the work must be satisfactory (Horibe, 1999; Osterloh & Frey, 2000). Osterloh and Frey (2000: 540) describe three major advantages of intrinsic motivation, which are also relevant to the events industry: first, intrinsic motivation can be more helpful than extrinsic in creative work; second, when people are intrinsically motivated, they will focus not only on the tasks they will be compensated for; and third, intrinsic motivation helps sharing tacit knowledge better than extrinsic motivation. This third aspect is especially valuable in event organisations, where, as previously discussed, a lot of tasks require tacit knowledge and embodied know-how. Von Krogh (2002: 89) similarly points out that;

> *if knowledge to be shared is tacit, the role of intrinsic motivation outweighs the role of extrinsic motivation, since no material incentive can change people's interest to codify and share tacit knowledge, and because no contract can assure effective and efficient knowledge sharing. [...] Hence, intrinsic motivation, by which team members realize immediate need satisfaction by working together with others in order to solve complex tasks, is a prerequisite for (tacit) knowledge sharing.*

Intrinsic and extrinsic motivation to share knowledge is, of course, also important at the group or team level. Mixed results have been published on which one is more beneficial for effectively sharing knowledge, but Hung et al. (2011: 425) for example, found that, "(…) economic reward, together with reciprocity and altruism, positively influence meeting satisfaction. [… and] that reputation feedback served as a strong incentive for both quantity and quality of knowledge shared." Similarly, Lam and Lambermont-Ford (2010) argued that financial extrinsic motivation does not appear to be relevant or enough on its own, but rather should be combined with other intrinsic incentives, or what the authors call 'hedonic' motivators. These derive from activities which are competence enhancing and enjoyable for employees, and are associated with physical and social wellbeing. In turn, hedonic motivation can stimulate creativity and innovation for these employees, as it leads to knowledge-seeking behaviour.

Lastly, another important motivation for sharing knowledge is that it increases respect from other members of the group. For most people being treated with respect and being recognised for their hard work is more valuable than monetary compensation. But there are more incentives that can be used to increase motivation for sharing knowledge, such as performance evaluations, work load reductions or a day off, and other premiums. These factors affect both the person transferring knowledge and the person receiving it, although it is usually more important for the sender than for the receiver (Szulanski, 2000). In event organisations, recognising the hard work of volunteers, for example, is considered very important in terms of motivating them to engage in knowledge practices with other staff members. Their level of motivation in general is also crucial in terms of retaining them and getting them to come back each year, which can to some extent help alleviate the issue of (re-)training them and of reinventing the wheel, as found by Clayton (2016). Similarly, Ragsdell et al. (2013) also demonstrated how motivation and pride in doing a good job were among the factors positively influencing and shaping knowledge transfer behaviour among volunteers, as well as trust in the management of the event and in the quality of project knowledge, as will be further discussed in the next section.

While motivation is a very important factor and provides a lot of advantages for managing knowledge, it can also hinder the process. Osterloh and Frey (2000) highlight two common disadvantages of relying on intrinsic motivation: first, intrinsic motivation is more difficult to change than extrinsic motivation, and it is not sure whether the outcome of the change is better. Second, not all aspects of intrinsic motivation are positive. Envy and dominance, for example, are also intrinsically motivated feelings that can easily arise in any organisation. For both effects it is important to set specific goals and treat members of the organisation with care. Showing them that dominance, for example, will create dissatisfaction in the long run can be a solution to the problem. It is clear to see that these questions around motivation are closely related to the culture of an organisation, and 'the way things are done.'

Trust

Similar to other elements discussed in this chapter, the vast majority of research on trust and knowledge management has so far mainly investigated trust as an important enabler for knowledge sharing; with much less research into trust and knowledge creation, knowledge use, and other knowledge management activities. In the wider knowledge management literature, it is commonly argued that a lack of trust between employees will inhibit the extent of knowledge and the way they share knowledge with each other. Trust is "the belief that people have about the likely behaviour of others, and the assumption that they will honour their obligations (not acting opportunistically)" (Hislop et al., 2018: 185). Or, according to Du Plessis (2006: 30), "[t]rust can be defined as the firm reliance on the integrity, ability, or character of a person or thing. Trust is a necessity for knowledge sharing. The depth and breadth of knowledge that will be shared between individuals will be determined by their levels of trust towards one another." An open organisational culture, as discussed above, also relies on transparency and trust between employees at all different levels, which in turn supports knowledge processes and practices. On the other hand, if there is a lack of transparency and trust, knowledge will most likely get stuck in silos, and knowledge sharing and re-use will also be hindered. An open and effective knowledge culture therefore requires both trust amongst employees and different areas of the organisation, as well as transparency, which needs to be initiated by the leaders and directors of the organisation (Du Plessis, 2006). The role of leadership in knowledge management, as briefly touched on in Chapter 5, is therefore important to refer to again here. For example, Koohang et al. (2017) recently found that effective leadership within organisations positively contributes to and enhances trust among employees, helps to promote the implementation of new knowledge management processes and strategies, and in turn enhances organisational performance, innovation and success.

Trust in itself is a multi-faceted concept and has been applied to knowledge management in many different ways. For the purpose of this book and in relation to event organisations, the distinction between two forms of trust are worth discussing further. McAllister (1995), Holste and Fields (2010), and Rutten et al. (2016), for example, highlight the difference between affect-based trust and cognition-based trust, and argue that a combination of both is required for effectively sharing and using knowledge in organisations. Affect-based trust is the kind of trust which is grounded in mutual care between employees and their concerns for each other. It is emotion-based, where employees share an emotional investment in each other's wellbeing. Cognition-based trust, on the other hand, refers to employees' reliability and competence and how these are perceived by co-workers. It is a rational decision, based on whether co-workers have a history of performing capably and with competence. The vast majority of studies in knowledge management found that high levels of trust lead to high levels of knowledge sharing. But Rutten et al. (2016) were able to further confirm that indeed, a lower level of trust leads to less knowledge sharing. The difference in effects of high and low levels of trust were, interestingly, overall larger for affect-based trust and for tacit knowledge sharing, than for cognition-based trust and explicit knowledge.

With most of the knowledge required to run a successful event being tacit in nature, in line with the findings above, event organisations will therefore also require much higher levels of both affective and cognitive trust among their employees. In event organisations, however, the two types of trust are difficult to establish within a short period of time and between staff members who have potentially never worked with each other before. Or even if they do have the required competencies and generally care for their co-workers, during the stressful time of the event, these relationships can be difficult to nurture and maintain. Furthermore, it also needs to be acknowledged that trust in relation to knowledge management always involves an element of risk, where one employee provides knowledge to another and thus makes themselves vulnerable. The other employee might act opportunistically and not provide any knowledge in return. In other words, trust is based on reciprocity and mutual benefit, or at least an expectation thereof (Hislop et al., 2018). This again, is a particular challenge in event organisations and can easily lead to issues of knowledge hiding or hoarding.

Collaboration and co-creation

Much like motivation and trust, effective collaboration among the team is also a necessary feature of knowledge management and tends to be higher in open organisational cultures; at the same time, it enhances social relationships between employees, which in turn positively contribute to their relational knowledge practices. For example, participatory or consensus-based decision-making in an organisation can foster collaboration, which in turn leads to knowledge sharing. Collaboration can then enhance both the co-creation of new knowledge as well as the effective transfer of knowledge within the organisation. Furthermore, a sense of belonging within an organisation and organisational loyalty also shape collaboration and co-creation, because when staff members feel included, they feel they are part of and care about the organisation. Collaboration and co-creation can therefore be seen as the final important cultural elements underpinning knowledge practices within an organisation, or as Nag et al. (2007: 842) suggested, we need to look at the "intersection of identity ('who we are'), knowledge ('what we know'), and practice ('what we do')." Effective collaboration and the co-creation of the event experience are thereby the practices of 'what we do' as an organisation, which is of course influenced by the organisational identity and culture, as well as individual, group and organisational knowledge brought together. The three elements are closely related and constantly reinforce each other.

As discussed in Chapter 5, there is furthermore a strong link between human resource management and knowledge management, and it has been widely argued that the two are related in terms of the effectiveness of team collaboration and culture (Yahya & Goh, 2002; Currie & Kerrin, 2003; Gorelick et al., 2004; Lepak & Snell, 2007). In other words, choosing the right people and learning 'how to' work together in order to achieve the organisation's aims and objectives, are crucial steps in developing collaboration and in co-creating the event experience. An event organisation that values relationships and the co-creation of new knowledge can furthermore enhance its creative output (Carlsen et al., 2010; Larson, 2011). Co-creation, by definition, is about interactions with others, working together and exchanging ideas; it is an active, creative and social process (Akhilesh, 2017). Conversations and meetings with a range of stakeholders, for example, can enable the co-creation of new know-how, both in terms of creative elements of the event, as well as process-oriented and operational tasks in making the event happen. Internal and external stakeholders bring a lot of expertise and experience to the table, which can be tapped into when there is a collaborative culture in place and stakeholders are willing to share their knowledge with the team (van Niekerk & Getz, 2019). This can go beyond the immediate event organisation and also include other partners and institutions in the network, such as within an event portfolio network (Antchak, Ziakas, & Getz, 2019).

Lastly, collaboration and co-creation are also an effective way of dealing with conflicts and resolving them in a way that benefits both parties (Jordan & Troth, 2002). There always are high and low points in any collaborative work relationship, which need to be accepted and utilised effectively, for example through taking chances with each other (Gregory, 2010). In any organisation, knowing 'how to' collaborate and work with others is therefore crucial in terms of dealing with conflict.

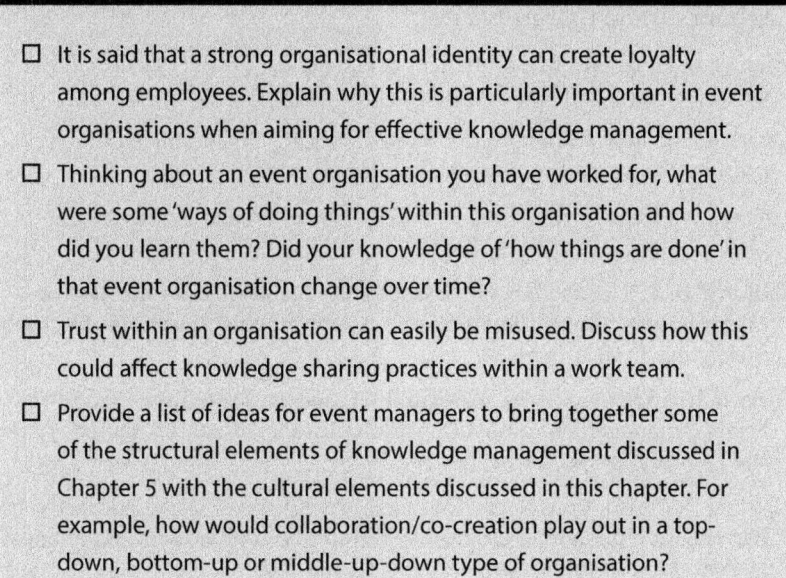

Study and discussion questions

- ☐ It is said that a strong organisational identity can create loyalty among employees. Explain why this is particularly important in event organisations when aiming for effective knowledge management.
- ☐ Thinking about an event organisation you have worked for, what were some 'ways of doing things' within this organisation and how did you learn them? Did your knowledge of 'how things are done' in that event organisation change over time?
- ☐ Trust within an organisation can easily be misused. Discuss how this could affect knowledge sharing practices within a work team.
- ☐ Provide a list of ideas for event managers to bring together some of the structural elements of knowledge management discussed in Chapter 5 with the cultural elements discussed in this chapter. For example, how would collaboration/co-creation play out in a top-down, bottom-up or middle-up-down type of organisation?

Recommended additional readings

Clayton, D. (2016). Volunteers' knowledge activities at UK music festivals: a hermeneutic-phenomenological exploration of individuals' experiences. *Journal of Knowledge Management, 20*(1), 162-180.

Du Plessis, M. (2006). *The Impact of Organisational Culture on Knowledge Management*. Oxford: Chandos Publishing.

Ragsdell, G., & Jepson, A. S. (2014). Knowledge sharing: insights from Campaign for Real Ale (CAMRA) festival volunteers. *International Journal of Event and Festival Management, 5*(3), 279-296.

References

Abfalter, D., Stadler, R., & Mueller, J. (2012). The organization of knowledge sharing at the Colorado Music Festival. *International Journal of Arts Management, 14*(3), 4-15.

Addleson, M. (2012). Will the real story of collaboration please stand up so we can see it properly? *Knowledge Management Research & Practice, 10*(4), 1-19.

Akhilesh, K. (2017). *Co-Creation and Learning: Concepts and Cases*. Springer.

Albert, S., & Whetten, D. A. (2004). Organizational identity. In M. J. Hatch & M. Schultz (Eds.), *Organizational Identity - A Reader* (pp. 89-118). Oxford: Oxford University Press.

Allen, J., O'Toole, W., McDonnell, I., & Harris, R. (2011). *Festival and Special Event Management* (5th ed.). Milton, Qld.: John Wiley & Sons.

Alvesson, M., & Sveningsson, S. (2008). *Changing Organizational Culture - Cultural change work in progress*. London: Routledge.

Antchak, V., Ziakas, V. & Getz, D. (2019). *Event Portfolio Management*. Oxford: Goodfellow Publishers.

Ashforth, B. E., Harrison, S. H., & Corley, K. G. (2008). Identification in organizations: An examination of four fundamental questions. *Journal of Management*, 34(3), 325-274.

Carlsen, J., Andersson, T. D., Ali-Knight, J., Jaeger, K., & Taylor, R. (2010). Festival management innovation and failure. *International Journal of Event and Festival Management*, 1(2), 120-131.

Clayton, D. (2016). Volunteers' knowledge activities at UK music festivals: a hermeneutic-phenomenological exploration of individuals' experiences. *Journal of Knowledge Management*, 20(1), 162-180.

Cruz, N. M., Pérez, V. M., & Cantero, C. T. (2009). The influence of employee motivation on knowledge transfer. *Journal of Knowledge Management*, 13(6), 478-490.

Currie, G., & Kerrin, M. (2003). Human resource management and knowledge management: Enhancing knowledge sharing in a pharmaceutical company. *International Journal of Human Resource Management*, 14(6), 1027-1045.

Deal, T. E., & Kennedy, A. A. (1982). *Corporate Cultures - The Rites and Rituals of Corporate Life*. Reading: Addison-Wesley Publishing Company.

Debowski, S. (2006). *Knowledge Management*. Milton, Qld: John Wiley & Sons.

Donate, M. J., & Guadamillas, F. (2011). Organizational factors to support knowledge management and innovation. *Journal of Knowledge Management*, 15(6), 890-914.

Du Plessis, M. (2006). *The Impact of Organisational Culture on Knowledge Management*. Oxford: Chandos Publishing.

Getz, D. (2002). Why festivals fail. *Event Management*, 7(4), 209–219.

Gorelick, C., Milton, N., & April, K. (2004). *Performance through Learning - Knowledge Management in Practice*. Amsterdam: Elsevier Butterworth-Heinemann.

Gregory, S. (2010). Collaborative approaches: Putting colour in a grey area. *International Journal of Community Music*, 3(3), 387-397.

Hecker, A. (2012). Knowledge beyond the individual? Making sense of a notion of collective knowledge in organization theory. *Organization Studies*, 33(3), 423-445.

Hislop, D. (2003). Linking human resource management and knowledge management via commitment. *Employee Relations*, 25(2), 182-202.

Hislop, D., Bosua, R., & Helms, R. (2018). *Knowledge Management in Organizations - A Critical Introduction* (4th ed.). Oxford: Oxford University Press.

Holste, J. S., & Fields, D. (2010). Trust and tacit knowledge sharing and use. *Journal of Knowledge Management*, 14(1), 128-140.

Horibe, F. (1999). *Managing knowledge workers: New skills and attitudes to unlock the intellectual capital in your organization*: John Wiley & Sons.

Hung, S.-Y., Durcikova, A., Lai, H.-M., & Lin, W.-M. (2011). The influence of intrinsic and extrinsic motivation on individuals' knowledge sharing behavior. *International Journal of Human-Computer Studies*, 69(6), 415-427.

Intezari, A., Taskin, N., & Pauleen, D. J. (2017). Looking beyond knowledge sharing: An integrative approach to knowledge management culture. *Journal of Knowledge Management*, 21(2), 492-515.

Jordan, P. J., & Troth, A. C. (2002). Emotional intelligence and conflict resolution: implications for human resource development. *Advances in Developing Human Resources*, 4(1), 62-79.

Kathiravelu, S. R., Mansor, N. N. A., Ramayah, T., & Idris, N. (2014). Why organisational culture drives knowledge sharing. *Procedia-Social and Behavioral Sciences*, 129, 119-126.

Kelly, D. (2000). Using vision to improve organisational communication. *Leadership & Organization Development Journal*, 21(2), 92-101.

Koohang, A., Paliszkiewicz, J., & Goluchowski, J. (2017). The impact of leadership on trust, knowledge management, and organizational performance. *Industrial Management & Data Systems*, 117(3), 521-537.

Lam, A., & Lambermont-Ford, J. P. (2010). Knowledge sharing in organisational contexts: a motivation-based perspective. *Journal of Knowledge Management*, 14(1), 51-66.

Larson, M. (2011). Innovation and creativity in festival organizations. *Journal of Hospitality Marketing & Management*, 20(3-4), 287-310.

Lennon, A., & Wollin, A. (2001). Learning organisations: empirically investigating metaphors. *Journal of Intellectual Capital*, 2(4), 410-422.

Lepak, D. P., & Snell, S. A. (2007). Managing the human resource for knowledge-based competition. In R. S. Schuler & S. E. Jackson (Eds.), *Strategic Human Resource Management* (2nd ed., pp. 333-351). Oxford: Blackwell Publishing.

McAllister, D. J. (1995). Affect-and cognition-based trust as foundations for interpersonal cooperation in organizations. *Academy of Management Journal*, 38(1), 24-59.

Muskat, B., & Deery, M. (2017). Knowledge transfer and organizational memory: an events perspective. *Event Management*, 21(4), 431-447.

Nag, R., Corley, K. G., & Gioia, D. A. (2007). The intersection of organizational identity, knowledge, and practice: Attempting strategic change via knowledge grafting. *Academy of Management Journal*, 50(4), 821-847.

O'Dell, C., & Hubert, C. (2011). Building a knowledge-sharing culture. *Journal for Quality and Participation*, 34(2), 22.

Orlikowski, W. J. (2002). Knowing in practice: Enacting a collective capability in distributed organizing. *Organization Science*, 13(3), 249-273.

Osterloh, M., & Frey, B. S. (2000). Motivation, knowledge transfer, and organizational form. *Organization Science*, 11(5), 538-550.

Ragsdell, G., & Jepson, A. S. (2014). Knowledge sharing: insights from Campaign for Real Ale (CAMRA) festival volunteers. *International Journal of Event and Festival Management*, 5(3), 279-296.

Ragsdell, G., Espinet, E. O., & Norris, M. (2013). Knowledge management in the voluntary sector: a focus on sharing project know-how and expertise. *Knowledge Management Research & Practice*, 12(4), 351-361.

Ram, M., & Holliday, R. (1993). Relative merits: family culture and kinship in small firms. *Sociology*, 27(4), 629-648.

Renzl, B. (2007). Language as a vehicle of knowing: the role of language and meaning in constructing knowledge. *Knowledge Management Research & Practice*, 5(1), 44-53.

Rutten, W., Blaas-Franken, J., & Martin, H. (2016). The impact of (low) trust on knowledge sharing. *Journal of Knowledge Management*, 20(2), 199-214.

Schein, E. H. (2004). *Organizational Culture and Leadership* (3rd ed.). Jossey-Bass.

Smith, K., & Lockstone, L. (2009). Involving and keeping event volunteers: Management insights from cultural festivals. In T. Baum, M. Deery, C. Hanlon, L. Lockstone, & K. Smith (Eds.), *People and Work in Events and Conventions - A Research Perspective* (pp. 154-170). London: CABI.

Stadler, R. (2013). Power relations and the production of new knowledge within a Queensland Music Festival community cultural development project. *Annals of Leisure Research*, 16(1), 87-102.

Stadler, R., & Fullagar, S. (2016). Appreciating formal and informal knowledge transfer practices within creative festival organizations. *Journal of Knowledge Management*, 20(1), 146-161.

Stadler, R., Fullagar, S., & Reid, S. (2014). The professionalization of festival organizations: A relational approach to knowledge management. *Event Management*, 18(1), 39-52.

Szulanski, G. (2000). The process of knowledge transfer: A diachronic analysis of stickiness. *Organizational Behavior and Human Decision Processes*, 82(1), 9-27.

Van der Wagen, L., & White, L. (2014). *Human Resource Management for the Event Industry* (2nd ed.). London and New York: Routledge.

Van Niekerk, M., & Getz, D. (2019). *Event Stakeholders*. Oxford: Goodfellow Publishers.

Von Krogh, G. (2002). The communal resource and information systems. *Journal of Strategic Information Systems*, 11(2), 85-107.

Yahya, S., & Goh, W.-K. (2002). Managing human resources toward achieving knowledge management. *Journal of Knowledge Management*, 6(5), 457-468.

7 Power and Knowledge

Learning objectives

- Explore power, politics and conflict in event organisations and their impact upon knowledge management.
- Learn to identify 'expert power' and 'legitimate power' within organisations and understand how they relate to knowledge management.
- Define power as a positive resource for knowledge to be created and shared, and explore opportunities for 'empowerment' within organisations.
- Understand the concept of power/knowledge and apply it to event organisations.

Introduction

This chapter discusses the links between knowledge and power and aims to demonstrate that, to "manage knowledge implies use of power, in terms of the ability of an organisation to achieve a collective sense of 'what to do next' and to exercise authority over the behaviour and communication patterns of internal and external agents – thereby influencing such things as who will interact with whom, on what basis, and to what purpose" (Clegg & Ray, 2003: 23). Power, knowledge, and the 'rules of the game' within an organisation (its political system, organisational culture, how things are done) are therefore intertwined; one cannot simply exist without the other. Clegg and Ray (2003: 23) go on to say, "the interaction of power and rules – to enable and constrain legitimate individual and collective actions – simultaneously shapes those actions. Rules shape actions that, in turn, have consequences for the evolution of rules and their interpretation in context."

It is important to note that based on the distinction between the objectivist and practice-based understanding of knowledge management used in this book, the question of power has also been dealt with rather differently in academic literature. Under the former, questions of power, politics and conflict are largely ignored. Power (as well as knowledge, for that matter) is simply regarded as something possessed by somebody 'over' someone else. It is a zero-sum game, where the saying 'knowledge is power' completely dominates the field. This can be based on, for example, power through hierarchical organisational structures, authority, as well as power through knowledge hoarding, and has been discussed by many scholars (see for example, Dixon, 1999; Willett, 2000; Pervaiz et al., 2001; Liebowitz, 2008). The first part of this chapter will explore the concept of power in this sense further. It will be clear to see that there is a lot of research on stakeholder power in events, but not much on power and knowledge, or power and knowledge management more specifically.

More recently, and in line with the practice-based understanding of power, it has been argued that power can also be regarded as a positive resource, in the sense that power and power relations can 'produce' new knowledge when people interact with each other and create a shared meaning (Gordon & Grant, 2005; Heisig, 2014; Heizman & Olsson, 2015; Stadler, 2019). The work of Foucault (1977, 1980, 1982) is crucial to discuss here, especially his concept of power/knowledge which shows that the two are inseparable. Nicolini (2007, 2011) further suggests that because knowledge practices are always collaborative in nature, inequalities are constantly produced and reproduced, which in turn leads to questions around power and conflict. "Power can [therefore] be seen as 'the rules of the game', which both enable and constrain action" (Clarke & Jepson, 2011: 9). This will be the subject for the latter sections in this chapter, where emphasis is put on how power and power / knowledge could be both a positive and a negative resource in event organisations, depending on how they are practised during the different stages of the event life cycle. If managed well, opportunities to 'empower' staff members, volunteers, and other stakeholders can arise and shape a very positive way forward when engaging in various knowledge practices.

Power, politics and conflict in organisations

Power, politics and conflict can, of course, arise in any organisation, even in the most egalitarian types of organisations, as well as between the organisation and other external partners or stakeholders. Organisations are, after all, political systems, where individuals, groups, teams, or departments constantly compete with each other for resources, such as money or space. A range of political tactics (for example, power games) can be played here to gain access to these resources. At the same time, it might occasionally be beneficial for individual employees to simply 'play the (political) game', in order to not create any conflict or dispute. In other words, politics within an organisation does not necessarily have to be a negative thing, it can also be constructive in leading change and innovation (Buchanan & Badham, 2020). In an events context, Larson and Wikström (2001), for instance, found that managing by consensus usually leads to stability, through for example, mutual commitment and trust; whereas conflict within the organisation highlights tensions, or power games between actors, which can – in a positive sense – create innovation and change. It is therefore important to acknowledge both, consensus and conflict, to coexist within organisations and in any relational interaction between co-workers, who in turn employ strategies for using one or the other at different times and for different purposes – sometimes intentionally, at other times unintentionally. In relation to knowledge management, it is therefore important to know and understand the political system, the rules of the game, and people's roles within this, in order to effectively engage in knowledge practices with others, rather than using power and politics in a destructive, anti-social way, or to play dirty tricks.

Surveillance is another example of how politics and power can play out in an organisation. Clegg (1989: 191) maintained that surveillance can occur in a personal, technical, bureaucratic or legal sense. "Its types may range through forms of, for instance, supervision, routinization, formalization, mechanization and legislation, which seek to effect increasing control of employees' behaviour, dispositions and embodiment, precisely because they are organization members." But it is important to understand that surveillance is not necessarily just about direct control. It could also be about cultural practices and norms, moral questions within the organisation, or formalised technical knowledge (Clegg, 1998). For example, many organisations use an open office design, which could be described as a political approach to utilise 'surveillance' techniques: employees might feel like they are being observed, watched or overheard by managers while they work with each other, take phone calls, or attend meetings. This could be seen as a negative approach of using power to get employees to work harder, as they will naturally compare themselves with each other and feel like they are being watched closely. It might also make them share only certain types of knowledge and not other types of knowledge, depending on what they want to achieve and how they want to present themselves. At the same time, however, it can also be an opportunity for employees to learn and understand 'how to' behave, express themselves (through observing others at work, for example), and 'how to' collaborate with each other. In terms of knowledge management, therefore, rather than using an open office design as a tool for surveillance, new knowledge can be produced by staff members watching each other work, and knowledge practices can be developed together, shared and used more effectively and efficiently than, for example, in silos.

Power as a resource

As mentioned at the beginning of this chapter, the traditional perspective on power regards it as a resource which can be used to influence other people in order to achieve a certain goal. In other words, power is used by one actor 'over' another. Along those lines, knowledge would then be regarded as a power resource which can be used to influence others and to achieve certain interests (Jepson & Clarke, 2014; Hislop et al., 2018). For example, if there is only one PR expert working for an event organisation, their specific PR knowledge will be a powerful resource available to them as well as to the organisation as a whole, and puts them in a high position of power (through their specialist knowledge). This is commonly referred to as 'expert power' and can strategically be used when trying to influence others: for example, the PR expert might choose to only share some of their knowledge and hoard other knowledge, or they might choose to share the knowledge only with a selected few colleagues or managers, but not with others. Furthermore, if their expert knowledge, skills and expertise are scarce in the industry as a whole, then they will become highly sought after, which again gives them a lot of power (e.g. when negotiating salaries, contracts, or specific job arrangements).

Thinking about power as a resource also often means that power can be embedded in the structure of the organisation. In many (particularly in hierarchical) organisations, power is structurally embedded in employment relationships. Hierarchies and structures imply a set of rules that need to be followed – some of them explicitly stated, others unspoken. According to Clegg (1989), it is those organisational structures and hence rules, that shape action, authority within the organisation, as well as control. Knowing about those rules is therefore important in terms of gaining an understanding of what is acceptable in the organisation and what not, how to behave, how to perform certain tasks. They constitute part of the important tacit knowledge that employees need to acquire. Front-line workers, for example, are usually in a subordinate position to management and hence possess less power to make decisions (Hislop et al., 2018). In an event organisation, seasonal staff members and volunteers might therefore be in a lower position of power than year-round, permanent staff, due to the fact that these permanent staff members tend to know more about the event and the organisation itself. In a festival context, for example, Clarke and Jepson (2011) explored how the Festival Steering Committee held most of the power over a community festival in Derby, and power was used as a source of discipline or control based to the steering committee's hierarchical position. It was through 'power brokers' that different definitions of 'culture' were used to include and exclude certain communities, and hence power was exercised over them.

Someone who has the legitimate right (e.g., through their higher hierarchical position) to influence others, is therefore said to have a lot of 'legitimate power.' If somebody has legitimate power, then others are most likely going to comply with what they are told to do. In terms of legitimate power, Reid (2011) for example, investigated event stakeholder power in rural and regional areas in Australia, and found that power was most commonly constituted through the length of time that different stakeholders had been involved with the event. It was this involvement over a long period of time that gave them legitimate power to make certain decisions. Furthermore, the largest stakeholder groups held significantly more power than smaller stakeholder groups.

Reid (2011) also noted that stakeholder power was exerted in three main ways: financial power (stakeholders that contributed to the event financially had more power), resource ownership (for example, the use of facilities provided by a key stakeholder) and influence (in terms of decision-making during the event planning process).

In relation to power through hierarchical structures, Batty (2016) found that within community sports event organisations in New Zealand, many stakeholders had legitimate power due to their position. However, in some cases pre-existing relationships between stakeholder groups (for example, because they had worked together in other contexts) meant that the local or host community perceived their actual level of power to be different to what the event organisers had assigned them in terms of their hierarchical position. In their study on power relations at Glastonbury Festival, White and Stadler (2018) also found that the local community felt they were not involved in any decision-making regarding the festival and did not share a common vision for the festival with the event organisers and other stakeholders. This lack of a shared vision allowed the organisers to exercise and misuse their power 'over' the local community, who at the same time and as a result of this, felt powerless and alienated. Similar issues around the use and misuse power by certain stakeholder groups (in most cases based on their hierarchical position, and to some extent also their level of knowledge due to this position) – both intentionally and unintentionally – are also highlighted by, for example, Jarman (2018), Kwiatkowski and Hjalager (2018) and Đurkin and Wise (2018).

Similarly, the knowledge management roles discussed in Chapter 5 are also shaped by these employment and power relations. Knowledge champions and strategists, for example, can have a lot of legitimate power, due to their (high) position within the organisation. But even in very flat organisational structures, or in communities-of-practice, where no hierarchy exists, not all members of the organisation or community are necessarily equal: as mentioned in Chapter 4, communities-of-practice have been critiqued for not taking power issues into account. For instance, one could argue that newcomers to the community will have less power than old-timers; or members of the core group of the community will have more knowledge and therefore more power than the active group and peripheral members. At the same time, members of the community might work together very collectively and collaboratively as long as they are within this community-of-practice, with no apparent power hierarchies, but within the wider organisation they might still be competing for promotions or other rewards. Taking these types of legitimate power into account is therefore paramount when engaging in knowledge practices, as they can shape the way employees behave, communicate with each other and make decisions.

Other important types of power to consider include reward, coercive and referent power. Table 7.1 below provides a summary of eight power bases as identified by French et al. (1959) and Benfari et al. (1986). It is important to note that, in relation to knowledge management, expert power (as already discussed above) can be regarded as the obvious example, with power being based on expert knowledge, whereas with other types of power, the link to knowledge and knowledge management is likely to be implicit or indirect. Coercive power, for example, is still to some extent based on employees' knowledge of how penalties could be administered, and in what way these penalties or sanctions might be unwelcome (e.g., hindering their progress).

Table 7.1: Eight different power bases. (Adapted from Benfari et al., 1986, and Buchanan & Badham, 2020)

Type of power	Description
Reward power	power based on the belief that in return for compliance a reward will be received
Coercive power	power based on the belief that someone can administer penalties or sanctions which would be unwelcome for the receiver
Referent power	power based on desirable abilities or personality traits which someone possesses and which can and should be copied by others
Legitimate power	power based on the belief that someone has authority to give directions within the boundaries of their position or rank
Expert power	power based on superior knowledge relevant to the situation and the task in hand
Information power	power through access to information which is not publicly available
Affiliation power	power through being associated with somebody who has authority, e.g. executive secretaries or personal assistants
Group power	power through problem-solving or conflict resolution in a group, where the group effort is greater than the contributions of individual members

All of the above discussions around power are only based on individuals or groups having power. At an organisational level however, opinions differ; some say individual knowledge workers may create, share, use, or develop knowledge in order to achieve organisational goals, but the knowledge is fundamentally theirs to use as, when and how they want. Whereas others say, it is actually the organisation that 'owns' the overall organisational knowledge created by all its employees, and therefore the organisation itself also has the power to manage this knowledge. This raises a question of who owns and controls the knowledge – knowledge workers, or the organisation they work for? – and often creates an interesting power relationship between the two (Hislop et al., 2018). Especially in the events industry, where some seasonal staff members rotate between different event organisations throughout the year, this can be a particular issue. They might take knowledge gained in one organisation and move on to another, where this knowledge can then be further implemented and used, which will potentially put that organisation in a better competitive position.

Lastly, when talking about power as a resource, more recent publications in event studies (see for example, Walters & Jepson, 2019a, 2019b; Walters et al., under review) point to yet another important issue around the use and misuse of power: in events, it is quite often the power of the privileged that means marginalised communities are misunderstood, underrepresented, or even entirely excluded from events. Marginalisation can be based on characteristics such as age, religious beliefs, sexuality, race, ethnicity, socio-economic status, disability, refugee or migrant status, or simply geographic location. Many event organisers use power relationships to adopt either inclusionary or exclusionary approaches to management practices, and consequently, power and knowledge across these events can be used either as a form of resistance on the one hand, or transformation on the other.

While the examples of event stakeholder power summarised throughout this section do not directly address the question of power and knowledge, or power and knowledge management, it can be concluded that there are two main issues that need further investigation in event organisations: First, in many cases, there seems to be a lack of knowledge about how to effectively engage stakeholders in the decision making and planning process. Event organisers therefore tend to use their knowledge as a power resource and hence use power over other stakeholders, rather than include them in the planning process and make their voices heard; and second, there is a lack of knowledge or understanding of other communities, of who they are and what is important to them, such as local and host communities, marginalised communities, or communities that are 'different' to what we consider to be the norm in society. Event organisers thus end up making assumptions, and again, exercise power *over* these communities, rather than use power in a more positive way in order to empower and transform. The relation between power and knowledge – or, perhaps in many cases, the misuse of power due to lack of knowledge – is therefore evident.

The concept of power/knowledge

More recent discussions around power highlight that rather than describing one person as having power over another, it is perhaps best to acknowledge that this power depends on the other person "acting in concert with what the first person does" (Rouse, 2006: 109). We can therefore think of power not as a resource, but rather as a complex network of social relations. These networks are not static, they are dynamic; and power therefore constantly changes and circulates within and between them (Foucault, 1980; Rouse, 2006). In terms of the hierarchical structures mentioned in the previous section, it is important to understand that power is employed at all levels and across many different dimensions, not merely top-down. Everybody in an organisation is constantly engaged in various power relations; managers do not simply have *all* the power *all* the time due to their hierarchical positions. In fact, Alvesson (1996: 64) pointed out that managers, "may be subordinate to cultural ideas and values which they take for granted. Intentions are not necessarily central. Power may thus be in force even if particular agents such as managers have limited control over it, and some of its functioning and effects are unintentional." Some of this power may be based on historical descriptions or stories of what the organisation is and what it aims to achieve. In turn, gaining the knowledge of how to express these stories, values, and ideals is an important part of understanding how certain power relations work in terms of knowledge management.

Asking the question, "who has power?" is therefore pointless; rather one should be asking, "how does power operate through practices, techniques or procedures?" (Townley, 1993). Similarly, the practice-based and relational understanding of knowledge acknowledges that knowledge is not static either; it is created and produced through statements, objects, practices, skills, social networks, organisations and institutions. This makes the relationship between power and knowledge an interesting, but very complex, topic to investigate. It is important to remember to not only think about 'knowledge is power', but to also look into 'power is knowledge', in order to better understand knowledge management and knowledge practices (Gordon & Grant, 2005) or as Foucault (1980: 52) famously stated: "It is not possible for power to be exercised without knowledge, it is impossible for knowledge not to engender power."

The concept of power/knowledge as defined and discussed by Foucault (1977, 1980, and 1982) highlights that power can never be a property that can be possessed. According to Foucault, power is a relation operating across all different levels of society. This power can be exercised within an organisation through, for example, discipline, through organising space or employees' activities, or even through stories about the success of the organisation, through rituals, photographs and other documents. In this way, employees will learn to live by certain rules and norms. In turn, their knowledge of what is normal and acceptable within the organisation, is implied within the power relations and either reinforces them or works against them, which means both power and knowledge are constantly negotiated and renegotiated.

In other words, power and knowledge are inseparable, but neither one is more important than the other. More knowledge does not necessarily mean more power, because employees can use power and knowledge, but they cannot ultimately possess either of them; rather, power/knowledge is embedded within social relationships and the ways employees work, or in the ways they talk. Acts of power are further embedded in the way we understand certain things (i.e., knowledge), while at the same time, expressing or using knowledge always involves the exercise of power (e.g. through prioritising one perspective over another, or through questioning one particular perspective) (Hislop et al., 2018). This is very much in line with the practice-based understanding of knowledge and knowledge management. Actor A would in this sense not exercise their power over actor B, but rather both actors A and B constitute their power relationship through the way they interact with each other. They are both equally important in the constitution of power, and hence, knowledge (Heiskala, 2001; Hislop et al., 2018).

To summarise, Foucault's concept of power/knowledge has two important implications: 1) Foucault challenges the traditional distinction between power and knowledge, where more knowledge can lead to more power, or where power can be enhanced through acquiring new knowledge. Instead he says, the two are attached to each other and occur side-by-side. 2) Power is not negative; it is creative in that it helps create new objects and therefore new knowledge.

Power and empowerment

Following on from this discussion of the concept of power/knowledge, using power in a positive sense within an organisation can help empower employees as well as other stakeholders. Empowerment can be defined as, "the granting of authority to employees to make key decisions within their area of responsibility" (Mosley et al., 2001: 37). The relationship between power and knowledge is again evident here: employees can only be granted authority, if they have the necessary knowledge to make these decisions. In turn, this authority can also mean that they are allowed to bend the rules, if necessary, because they have the knowledge and expertise to judge the circumstances and act accordingly. Through empowering employees, a climate and culture of trust can then be established – key factors for effectively practising knowledge management, as discussed in Chapter 6. Empowered employees feel a sense of commitment and ownership, a sense of responsibility, and become more knowledgeable about the job and its wider implications for the organisation as a whole. This positive use of power, in the sense that it is redistributed among all members of the organisation, is an excellent way of enhancing knowledge practices, both vertically as well as horizontally (Mosley et al., 2001).

More specifically, Spreitzer (2008) summarised 20 years of research on empowerment and distinguishes between social-structural and psychological empowerment (see Figure 7.1). A combination of both is ideal within an organisation: Social-structural empowerment "focuses on how organizational, institutional, social, economic, political, and cultural forces can root out the conditions that foster powerlessness in the workplace" (Spreitzer, 2008: 55). These include, for example, participatory decision-making, skill and/or knowledge-based pay, open flow of information, flat organisational structure, and training. In other words, it is those elements that are important to implement at an organisational level. All of them are, of course, closely related to knowledge management and to performing and engaging in knowledge practices. The social-structural perspective on empowerment does not, however, take into account how individual employees experience this sense of empowerment. This is where the second type of empowerment – psychological empowerment – is equally as important to consider, which refers to "a set of psychological states that are necessary for individuals to feel a sense of control in relation to their work" (Spreitzer, 2008: 56).

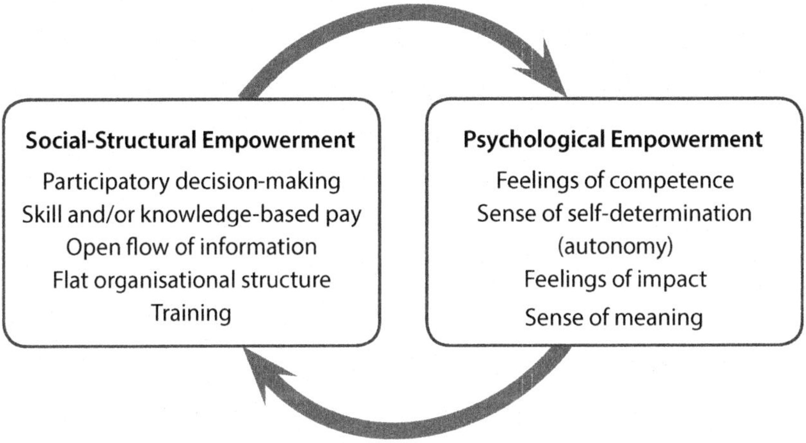

Social-Structural Empowerment
Participatory decision-making
Skill and/or knowledge-based pay
Open flow of information
Flat organisational structure
Training

Psychological Empowerment
Feelings of competence
Sense of self-determination (autonomy)
Feelings of impact
Sense of meaning

Figure 7.1: Types of empowerment

Whilst not explicitly stated by Spreitzer (2008), the links to knowledge management are also evident in the psychological empowerment theory. The different dimensions include:

- *Feelings of competence*: an employee feels they are able to successfully perform their tasks; they feel they have the knowledge and skills to do the job;

- *Sense of self-determination (autonomy)*: autonomous employees feel they have the necessary freedom and independence to decide which activities they need to engage in to perform their job;

- *Feelings of impact*: employees feel empowered if they feel they can impact the results of administrative, operational and/or other elements of the job; and

- *Sense of meaning*: employees get a sense of meaning from their job, if it fits with their career goals, beliefs, values and behaviours.

Surprisingly, in an events context, and in terms of event management in particular, not much has been written about empowerment thus far. It is important to note that through empowerment, employees can feel more committed to their jobs and are hence more likely to stay with the organisation – a crucial factor in the events industry where retaining employees can be a challenge, as previously discussed in this book. Moreover, empowerment can also take place between the event organisation and other stakeholders, such as the local or host community, or the marginalised communities that often feel powerless if traditional ways of working and managing the event experience are employed. For an example of how the power/knowledge relation can be applied to a festival context in a positive sense and how a community can be empowered through this experience and knowledge, see Stadler (2013). In the Queensland Music Festival organisation investigated in this study, new forms of knowledge were created by shifting power relations between the organisation and the community. The event managers and production team did not simply impose their ideas upon the community, but rather engaged them in a community cultural development project in an attempt to make the community's voices heard. This in turn, meant that the festival organisation needed to constantly learn and relearn new things, change their ways of thinking and working, and hence create new knowledge for the organisation along the way – together with the community, who were seen as equal partners in the project. The result was a very empowering experience for the community, who may not have much power in a traditional sense. The newly gained sense of empowerment was a very positive outcome of the effective use of power and knowledge, while also creating new knowledge for the organisation itself. The next case study provides more examples of knowledge and empowerment in a festival organisation.

It is clear to see that different approaches to power as discussed in this chapter can have very different implications for knowledge management. In a negative sense, power can be misused; one actor can use their power over another. Or, somebody with expert or legitimate power can, for example, deliberately hide or hoard their knowledge. Yet, in a positive sense, power can also be a resource for the creation of new knowledge and for different ways of working and thinking. A sense of empowerment can be a very beneficial factor for knowledge management and for all stakeholders involved, if managed carefully and appropriately.

Case study: Power, knowledge and empowerment at the Queensland Music Festival

Questions of power and empowerment in relation to knowledge management were never explicitly discussed amongst staff members at the Queensland Music Festival themselves (for an introduction to the case study, see Chapter 4). However, when prompted to talk about positive examples of knowledge management, these became evident in discussions with me, the researcher. They were discussed in different ways by senior staff members, seasonal staff and volunteers. But at the same time, common themes emerged around empowerment and how important this is in terms of bringing out the best in people, enhancing knowledge sharing practices and taking ownership of decisions made. Examples of interview quotes are presented below with the aim of bringing together the different perspectives: quotes from senior staff members are set to the left, and quotes from seasonal staff and volunteers to the right. (All names have been changed.)

Members of senior management team

Ultimately, the knowledge management issue for me is ensuring that the delegation of authority works. I think it's a really important principle that people have control of the projects both artistically/creatively and financially. And you have to let go and just let them get on with it.

As you can see it's a great environment to work in here. And we don't like them working late at night, and we recognise it has to happen, but we encourage them to go... we don't worry about if they wonder in at ten o'clock in the morning, or if they go off and do another job. I know Andy works at [a different venue] tonight.

That's fine, he doesn't take leave for that, I trust that he'll still get his job done, while he earns some extra money over there... and also there is a huge benefit in these people going off and doing different festivals as well, because then you learn other things! And you see what their pressures are and it helps picking up and learning from working other things, because it helps the festival. You've got to allow them to do that and trust that they will a) still get their jobs done, and b) bring back their experiences elsewhere and share them with everyone. But it's THEIR decision to make, some of them want to do that, others don't.

Seasonal staff members and volunteers

You're not supposed to know everything as a manager and it's a pretty harsh industry for that. You kind of assume that you do know everything. That's not correct! And hence why they hire people here that know different areas, and have different skills that complement each other, because you don't know everything. And then they trust you to get on with it and that's how it should be! Makes you feel quite good that you are allowed to make your own decisions because they trust you to know what you're doing. Otherwise they wouldn't have hired you in the first place.

The key principles for me really are: get clarity, use the right people, continue to trust them... the thing that works best for me is the engagement where you have the best possible people on each project, phenomenal artists, phenomenal producers, technical managers, marketing, who really deliver. And there is never any doubt that they won't deliver.

I did know a few of the staff here beforehand. And look, a lot of them work from festival to festival, so they all really know each other. I've met a few of them, probably 50% of them, before. So I knew they were great people and I also knew that they were really talented. So it was kind of good to be in that atmosphere where everybody, you know, is somebody you wanna be around. There is no personalities that don't work. They all support each other and understand each other. It's a good, strong team and everyone has got a very good work ethic. It's kind of one of those rare places where you walk in and if you get your job done, you'll be able to help someone else. They may need you, and vice versa.

I think, what's good about the team process here is that people share ideas and knowledge and out of that process, you get these little nuggets of gold that turn into something like this massive project we are working on just now!

In a festival you can have the best creative ideas, but we can only generate as much as we can generate with imagination or whatever. But feet on ground and people who are actually capable of seeing those things into a reality, are those people out there [points at the rest of the team]. And you have to be friendly! I mean, I am friendly anyway [laughs]. I can't see how you could possibly get the best out of anyone if you were aloof and uncommunicative. Anyway, it's never been my way… I also tend to be fairly upfront with problems or issues. Confronting situations… maybe a little bit too upfront sometimes [laughs]… But I think everybody has different roles and all of those roles are equally as important in making the festival work. Because if one of them doesn't work, you don't have a festival. So I think that everybody needs to be respected for their abilities that they bring to the enterprise. And I just think, that's kind of the basis.

> Here, you can walk out, ask a question, everybody knows the answer… it's just, everybody starts talking. And that's what it should be like! Let's all be friends here because we all have to work together. And… and it's a great thing. For example, the relationship between Steve and I is very good, because we are both so open. We both love what we do. We both love talking about things… there's things I don't necessarily agree with … with what he does in terms of marketing. And there's probably things, ideas, where he's probably going, "oh…" [rolls her eyes] but, that's just the way it is. But in the end, we put our ideas together and make a decision and go with it. And it's great working with Steve, because I can do that. My last boss at [a different festival], she was sort of… I tell her an idea and it somehow became HER idea in the end, it was sort of weird. I didn't like that. If I don't have a good relationship with my boss, I tend to keep things to myself… I'm not going to tell you anything. But here, it's like the opposite. You can say and share everything with everyone, nobody is going to steal your ideas. If anything they pat you on the back and make you feel good about it. Congratulate you on your great idea.

> You hear stories about other organisation where there is a cultural secrecy and knowledge is power. And I just go like, "what's the point?! It doesn't help anybody!" If you want to bring people on to help you solve problems, you've got to share the information. Where it's gone wrong this year for us, I think, sometimes there is just not enough information to share, you know? When you don't know what the projects really are, then they will struggle. So that wasn't an intentional knowledge is power thing, but it was a sort of accidental knowledge is power thing of, well it's all in my head, why can't you understand it?! [laughs] Well, we can't understand it, because we don't get it, we haven't been told. So that's an interesting aspect of knowledge in the organisation, I'd say... So we need to manage that better.

>> Yeah, there's very few things that you ask here and people don't have an answer for or that's not your business to know. We keep a lot of records for previous shows and all the producers, when I ask a question, are quite comfortable explaining it. I never feel like they don't want to share certain things with me or they don't want me to know about something. I guess they know that at the end of the day, if I have that piece of information, I will be able to make better decisions too, sort of within my area of responsibility, obviously.

It is evident to see that senior staff at QMF were keen to empower their employees by sharing relevant information with them, respecting them and trusting them to make the right decisions – all elements of social-structural empowerment as outlined throughout this chapter. Seasonal staff members at the same time, expressed feelings of empowerment not only in terms of decision-making but also in working with each other. Rather than competing within their teams, there was a sense of collegiality and an acknowledgement of everyone's skills, knowledge and expertise. They felt they had the autonomy to make decisions within their areas of responsibility, they felt a positive impact from that, as well as a sense of meaning and feelings of competence – all previously discussed elements of psychological empowerment. In turn, senior managers were aware of the importance of hiring people who will work well together in the first place and then bringing out the best in them through sharing relevant knowledge with them. It is clear to see that both, social-structural and psychological empowerment are hence needed in order to create an

> empowering organisation overall. On both sides, comparisons to other festival organisations were made in terms of openness, trust and empowerment versus misuses of power through hierarchy, secrecy and domination. This explicit and shared understanding of 'how things should be' helped create a knowledge-friendly and very empowering culture at QMF.
>
> (Based on research by Stadler, 2013)

Study and discussion questions

- Discuss in your own words how 'power' can be both a positive and a negative resource for knowledge management.
- Why do you think questions of power, politics and conflict are largely ignored in the wider knowledge management literature?
- Review the eight different power bases mentioned in this chapter and discuss examples of each in an event organisation.
- Based on your own experience of organising events, provide some practical examples of how volunteers can be empowered when sharing their knowledge
- Work through the interview quotes in the case study and identify examples of the most important elements of empowerment as discussed throughout the chapter and particularly in Figure 7.1

Recommended additional readings

Batty, R. J. (2016). Understanding stakeholder status and legitimate power exertion within community sport events: A case study of the Christchurch (New Zealand) City to Surf. In A. Jepson & A. Clarke (Eds.), *Managing and Developing Communities, Festivals and Events* (pp. 103-119). London: Palgrave Macmillan UK.

Jepson, A., & Clarke, A. (2014). The future power of decision making in community festivals. In I. Yeoman, M. Robertson, U. McMahon-Beattie, E. Backer, & K. A. Smith (Eds.), *The Future of Events and Festivals* (pp. 67-83). Abingdon: Routledge.

Stadler, R. (2013). Power relations and the production of new knowledge within a Queensland Music Festival community cultural development project. *Annals of Leisure Research*, 16(1), 87-102.

References

Alvesson, M. (1996). *Communication, Power and Organization*. New York: Walter de Gruyter.

Batty, R. J. (2016). Understanding stakeholder status and legitimate power exertion within community sport events: A case study of the Christchurch (New Zealand) City to Surf. In A. Jepson & A. Clarke (Eds.), *Managing and Developing Communities, Festivals and Events* (pp. 103-119). London: Palgrave Macmillan UK.

Benfari, R.C., Wilkinson, H.E., & Orth, C.D. (1986). The effective use of power. *Business Horizons*, 29(3), 12-16.

Buchanan, D., & Badham, R. (2020). *Power, Politics, and Organizational Change*. Los Angeles: SAGE Publications.

Clarke, A., & Jepson, A. (2011). Power and hegemony within a community festival. *International Journal of Event and Festival Management*, 2(1), 7-19.

Clegg, S. (1989). *Frameworks of Power*. London: SAGE Publications.

Clegg, S. (1998). Foucault, power and organizations. In A. McKinlay & K. Starkey (Eds.), *Foucault, Management and Organization Theory - From Panopticon to Technologies of Self* (pp. 29-48). London: SAGE Publications.

Clegg, S., & Ray, T. (2003). Power, rules of the game and the limits to knowledge management: Lessons from Japan and Anglo-Saxon alarms. *Prometheus*, 21(1), 23-40.

Dixon, N.M. (1999). *The Organizational Learning Cycle - How We Can Learn Collectively* (2nd ed.). Hampshire: Gower.

Đurkin, J., & Wise, N. (2018). Managing community stakeholders in rural areas: Assessing the organisation of local sports events in Gorski kotar, Croatia. In A. Jepson & A. Clarke (Eds.), *Power, Construction and Meaning in Festivals* (pp. 185-200). London: Routledge.

Foucault, M. (1977). *Discipline and Punish - The Birth of the Prison*. New York: Pantheon Books.

Foucault, M. (1980). *Power/Knowledge - Selected Interviews & Other Writings 1972-1977* (edited by Colin Gordon). New York: Pantheon Books.

Foucault, M. (1982). The subject and power. *Critical Inquiry*, 8(4), 777-795.

French, J. R., Raven, B., & Cartwright, D. (1959). The bases of social power. *Classics of Organization Theory*, 7, 311-320.

Gordon, R., & Grant, D. (2005). Knowledge management or management of knowledge? Why people interested in knowledge management need to consider Foucault and the construct of power. *Journal of Critical Postmodern Organization Science*, 3(2), 27-38.

Heisig, P. (2014). *Knowledge Management - Advancements and Future Research Needs*. Paper presented at the British Academy Of Management Conference, Belfast.

Heiskala, R. (2001). Theorizing power: Weber, Parsons, Foucault and Neostructuralism. *Social Science Information*, 40(2), 241-264.

Heizman, H., & Olsson, M.R. (2015). Power matters: the importance of Foucault's power/knowledge as a conceptual lens in KM research and practice. *Journal of Knowledge Management*, 19(4), 756-769.

Hislop, D., Bosua, R., & Helms, R. (2018). *Knowledge Management in Organizations - A Critical Introduction* (4th ed.). Oxford: Oxford University Press.

Jarman, D. (2018). Personal networks in festival, event and creative communities: perceptions, connections and collaborations. In A. Jepson & A. Clarke (Eds.), *Power, Construction and Meaning in Festivals* (pp. 65-89). London: Routledge.

Jepson, A., & Clarke, A. (2014). The future power of decision making in community festivals. In I. Yeoman, M. Robertson, U. McMahon-Beattie, E. Backer, & K. A. Smith (Eds.), *The Future of Events and Festivals* (pp. 67-83). Abingdon: Routledge.

Kwiatkowski, G., & Hjalager, A.-M. (2018). Innovation in rural festivals: Are festival managers disempowered? In A. Jepson & A. Clarke (Eds.), *Power, Construction and Meaning in Festivals* (pp. 91-107). London: Routledge.

Larson, M., & Wikström, E. (2001). Organizing events: Managing conflict and consensus in a political market square. *Event Management*, 7(1), 51-65.

Liebowitz, J. (2008). 'Think of others' in knowledge management: making culture work for you. *Knowledge Management Research & Practice*, 6(1), 47-51.

Mosley, D.C., Megginson, L.C., & Pietri, P.H. (2001). *Supervisory Management - The Art of Empowering and Developing People* (5th ed.). South Western: Thomson Learning.

Nicolini, D. (2007). Stretching out and expanding work practices in time and space: The case of telemedicine. *Human Relations*, 60(6), 889-920.

Nicolini, D. (2011). Practice as the site of knowing: Insights from the field of telemedicine. *Organization Science*, 22(3), 602-620.

Pervaiz, K A., Lim, K.K., & Loh, A.Y.E. (2001). *Learning through Knowledge Management*. Oxford: Butterworth Heinemann.

Reid, S. (2011). Event stakeholder management: developing sustainable rural event practices. *International Journal of Event and Festival Management*, 2(1), 20-36.

Rouse, J. (2006). Power/Knowledge. In G. Gutting (Ed.), *The Cambridge Companion to Foucault* (2nd ed., pp. 95-122). Cambridge University Press.

Spreitzer, G.M. (2008). Taking stock: A review of more than twenty years of research on empowerment at work. In J. Barling & C. L. Cooper (Eds.), *The SAGE Handbook of Organizational Behavior: Volume One: Micro Approaches* (pp. 54-72). Los Angeles: SAGE.

Stadler, R. (2013). Power relations and the production of new knowledge within a Queensland Music Festival community cultural development project. *Annals of Leisure Research*, 16(1), 87-102.

Stadler, R. (2019). Knowledge management in event and festival organisations: Challenges and future directions. In E. Lundberg, J. Armbrecht, & T. Andersson (Eds.), *A Research Agenda for Event Management* (pp. 154-169). Cheltenham: Edward Elgar.

Townley, B. (1993). Foucault, power/knowledge and its relevance for human resource management. *Academy of Management*, 18(3), 518-545.

Walters, T., & Jepson, A. S. (2019a). Understanding the nexus of marginalisation and events. In T. Walters & A. S. Jepson (Eds.), *Marginalisation and Events* (pp. 1-16). Abingdon: Routledge.

Walters, T., & Jepson, A. S. (Eds.). (2019b). *Marginalisation and Events*. Abingdon: Routledge.

Walters, T., Stadler, R., & Jepson, A. (under review). Positive power: Events as temporary sites of power which 'empower' marginalised groups. *International Journal of Contemporary Hospitality Management*.

White, Z., & Stadler, R. (2018). "I don't think they give a monkey's about me" - Exploring stakeholder power and community alienation at Glastonbury Festival. In A. Jepson & A. Clarke (Eds.), *Power, Construction, and Meaning, in Communities, Festivals and Events* (pp. 21-34). Abingdon: Routledge.

Willett, C. (2000). Knowledge sharing shifts the power paradigm. In D. Morey, M. Maybury, & B. Thuraisingham (Eds.), *Knowledge Management - Classic and Contemporary Works* (pp. 249-259). Cambridge: The MIT Press.

8 Appreciative Sharing of Knowledge

> **Learning objectives**
>
> ☐ Define Appreciative Inquiry and discuss how the Appreciative Sharing of Knowledge approach is different to problem-solving approaches to knowledge management.
> ☐ Understand the importance of stories and positive language in Appreciative Sharing of Knowledge.
> ☐ Apply the four steps of Appreciative Sharing of Knowledge to event examples.

Introduction

The previous chapters in this book have so far mainly focused on problem-solving approaches to knowledge management within event organisations. It has been argued that there are a lot of knowledge management challenges in event organisations, which need to be overcome in order for the organisation to be successful in the long run. This chapter presents an entirely different approach to knowledge management: it introduces Appreciative Inquiry as an approach to management based on an organisation's strengths with regards to knowledge management, such as, for example, knowledge creation and knowledge sharing practices that are already working well. Applying the principles of Appreciative Inquiry and Appreciative Sharing of Knowledge, as defined by Cooperrider and Srivastva (1987), Cooperrider and Whitney (1999) and Thatchenkery and Chowdhry (2007), the aim is to utilise the strengths within the organisation, and learn from and further build on these strengths, in order to enhance the organisation's knowledge management practices and ultimately its success overall.

The first two sections of this chapter introduce Appreciative Inquiry and Appreciative Sharing of Knowledge as alternative approaches to knowledge management. They highlight key principles of these and provide examples as to how they can be applied to event organisations. The final section of the chapter more specifically discusses (positive) stories and storytelling within the Appreciative Sharing of Knowledge approach and focuses on how stories of success, achievement, and positive memories can be a useful tool within event organisations to create a shared understanding and knowledge of what the event is about, what it aims to achieve, and how to work together effectively and efficiently. It will be reemphasised that these tacit knowledge practices are invaluable within any organisation, and can provide a competitive advantage in the long term.

Appreciative Inquiry – making visible what works well in an organisation

Most of the management, knowledge management, as well as event management literature to date has emphasised problem-solving approaches. For example, previous chapters in this book have highlighted challenges and issues like knowledge hoarding, lack of trust between staff members, the inexperience of volunteers, or the misuse of power. An Appreciative Inquiry approach turns all of this upside down and identifies an organisation's strengths; it highlights everything that already works well within the organisation, in order to build on these strengths over time. This includes individual employees' strengths, as well as strengths developed at a group or team level and within the organisation as a whole. This is not easy to do, however, as Thatchenkery and Chowdhry (2007: 33) argued, "Being appreciative is harder than finding problems. To be appreciative, we must experience a situation, accept the situation, make sense of the situation (pros/cons), and do a bit of mental gymnastics to understand the situation with an appreciative lens. Not only that, the appreciative lens that we put on the situation impacts our next experience as well."

Appreciative Inquiry was first developed by Cooperrider and Srivastva and further developed by some of their colleagues (see for example, Cooperrider & Srivastva, 1987; Cooperrider et al., 1995; Whitney & Trosten-Bloom, 2003; Van Tiem & Rosenzweig, 2006). Their research mainly looked at organisational behaviour and initially explored both the strengths and weaknesses of a case study organisation. It was, however, particularly the strengths that they were fascinated by, as well as the many positive stories shared within the organisation, and they decided to develop these ideas further. The work is based on social constructionist underpinnings and developed these key principles of Appreciative Inquiry (summarised from Cooperrider & Whitney, 1999):

- *The constructionist principle*: The organisation needs to be understood as a living, human construction where relationships are the locus of knowledge, and the world is made sense of through the power of language;
- *The simultaneity principle*: Inquiry and change occur at the same time;
- *The poetic principle*: The organisation's story is co-authored by all its members; stories are sources for learning and interpretation;
- *The anticipatory principle*: By creating positive images of the future, current behaviours and actions are positive too; and
- *The positive principle*: Positive questions and stories provide momentum for change.

Based on these core principles, Appreciative Inquiry within the wider management literature, is defined as a step-by-step process to identify 'what is', 'what might be', 'what could be' and finally 'what will be' (Thatchenkery & Chowdhry, 2007), as can be seen Figure 8.1. However, the process does not always evolve sequentially across these different stages and not necessarily one at a time; it can be quite improvisational. Appreciative Inquiry will be different for every organisation, for every team, or community; and it continuously changes based on what is important to members of the organisation (Finegold et al., 2002; Whitney & Trosten-Bloom, 2003). The approach has been widely applied across different organisational settings as well as in tourism and hospitality research (Maier, 2008; Raymond & Hall, 2008; Koster & Lemelin, 2009), but research in event management using Appreciative Inquiry is scarce.

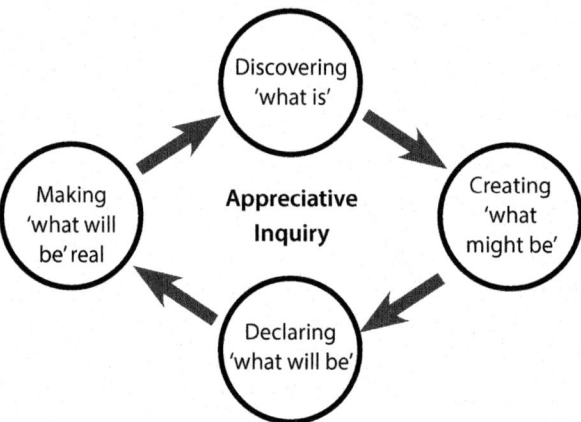

Figure 8.1: The four steps in Appreciative Inquiry

Furthermore, Appreciative Inquiry is a very collaborative approach, where as many employees as possible are encouraged to participate (Yoder, 2004). It is based on the organisation's existing culture and core values, which should ideally be shared by all employees. But most importantly, Appreciative Inquiry uses the best (past and present) successes of the organisation and builds on these: it affirms, reaffirms and makes explicit what is already working well within the organisation (Whitney & Trosten-Bloom, 2003; Van Tiem & Rosenzweig, 2006; Thatchenkery & Chowdhry, 2007). In other words, the aim is to appreciate what 'is', while at the same time envision what 'could be' possible for the organisation in the future (Rogers & Fraser, 2003). For example, an event organisation might identify that they are currently doing very well in working with their sponsors: they engage a range of sponsors and partners in decision-making, consult them along the way, and create impact for them as well as for the organisation. They have managed to create a mutually beneficial relationship with all sponsors. Appreciating these successes and identifying them as key to the success of the organisation, will help specify the future direction of the event. This could, for example, include building further partnerships with new sponsors by using a similar approach to negotiating a deal, keeping them engaged, and building a long-term relationship with them. It could also mean using similar strategies (that have been identified as successful strategies that 'work well') when working with other stakeholders, such as the local council, the local community, and the media. Knowledge created in one area of the organisation, could be transferred into another area and successful knowledge practices of how to work with sponsors could be applied to other stakeholders in the future in order to develop a more consistent approach of working with all stakeholders across all areas of the organisation.

It is important to note, however, that problems within an organisation are not completely ignored when using an Appreciative Inquiry approach. Rather, by focusing on the organisation's strengths and successes, problems can be turned into opportunities for learning (Cooperrider & Whitney, 1999; Thatchenkery & Chowdhry, 2007). Similarly, conflict can still arise when going through the four steps of Appreciative Inquiry; but rather than analysing the conflicts and problems and aiming to resolve them, they are turned into opportunities – for example, opportunities for learning, collaboration and reflexive thinking (Whitney & Trosten-Bloom, 2003; Van der Haar & Hosking, 2004). Referring back to questions around power and empowerment as discussed in the previous chapter, Appreciative Inquiry can therefore involve a very positive and productive use of power to create new knowledge and new ways of thinking.

Lastly, it has already been shown that Appreciative Inquiry can be a useful approach for the organisation as a whole. It can, however, also be applied at a group or team level, and even for individual employees. For example, it can help discover what "enables a particular group to work at its best" (Thatchenkery & Chowdhry, 2007: 44). Sharing best practices, successes and positive experiences within a team can help team members learn about each other's skills, knowledge and expertise, while at the same time create a shared understanding and image of what the team is and what is important to them. This constitutes, once again, important tacit knowledge that enhances relationships within the team, trust, and collaboration. It is particularly valuable when it comes to highly complex tasks and work activities, where an emphasis on what is important and useful for the task can help the team work through the more complex elements of it quickly and efficiently. In event organisations, there are of course many complex tasks that each individual team needs to accomplish, but the teams might have very different ways of approaching their tasks. For example, the marketing team might prefer to have regular, but very informal meetings to allow for creative ways of thinking and working to develop; the finance team might prefer a weekly meeting to discuss key updates, but only communicate via email throughout the rest of the week in order to have a written account of everything available, if needed at a later stage; and the technical team might implement a 'buddy' system when visiting the site and double-checking important developments and decisions with each other. All these different ways of working can be identified by each team as practices that allow them to 'work at their best'. Through the use of an Appreciative Inquiry lens, they can be made more explicit and used even more effectively in the future. Focusing on positive experiences and opportunities is regarded as important because "people experiencing positive feelings are more flexible, creative, integrative, open to information and efficient in their thinking" (Bushe, 2007: 32), qualities that are particularly valuable in event management.

Appreciative Sharing of Knowledge: what makes people share knowledge?

Based on the ideas of Appreciative Inquiry, Thatchenkery and Chowdhry (2007) later applied the same principles to knowledge management (and knowledge sharing in particular) and developed the Appreciative Sharing of Knowledge approach. In line with the original ideas, Appreciative Sharing of Knowledge is a prospective rather than retrospective approach. It regards knowledge management as something to embrace and aspire to, or in other words, one of many opportunities within an organisation contributing to success. It is very much centred around the question, 'what makes people share knowledge?', aims to identify knowledge sharing practices that already work well within the organisation, and based on these develop best practices for sharing in the future. According to Thatchenkery and Chowdhry (2007), communication and dialogue are key factors in the process. Through dialogue and communication, a shared vision of the future can be co-created between as many employees and other stakeholders as possible. In an event organisation that includes dialogue between permanent and seasonal staff, volunteers, contractors, sponsors, the local community, and many other stakeholders (van Niekerk & Getz, 2019).

Furthermore, any process of Appreciative Sharing of Knowledge is embedded within the current climate and culture of an organisation; in fact, "Appreciative Sharing of Knowledge usually exists in some form in many organisations, even though it is not known as such" (Thatchenkery & Chowdhry, 2007: 154). An open and collaborative organisational culture as discussed in Chapter 6 is once again very important here, as it enhances knowledge sharing and transfer not only within each team or group, but also across different levels and departments of the organisation. In line with Appreciative Sharing of Knowledge, employees can identify successful practices of creating knowledge in relation to each other, through conversation, communication and social interaction (Finegold et al., 2002). They can focus on learning from and with each other, break down hierarchies and silos, renegotiate power relations, and many other elements of knowledge management already discussed in this book, particularly relational knowledge practices. It is important to reiterate that there is not necessarily just one perfect knowledge practice that works across the entire organisation, but different and multiple local practices might emerge (between different teams or groups, for example) and successfully co-exist (Van der Haar & Hosking, 2004).

Based on the original steps of Appreciative Inquiry (Figure 8.1), the cycle of Appreciative Sharing of Knowledge as identified by Thatchenkery and Chowdhry (2007: 50) also consists of four key steps (see Figure 8.2 and further discussion below):

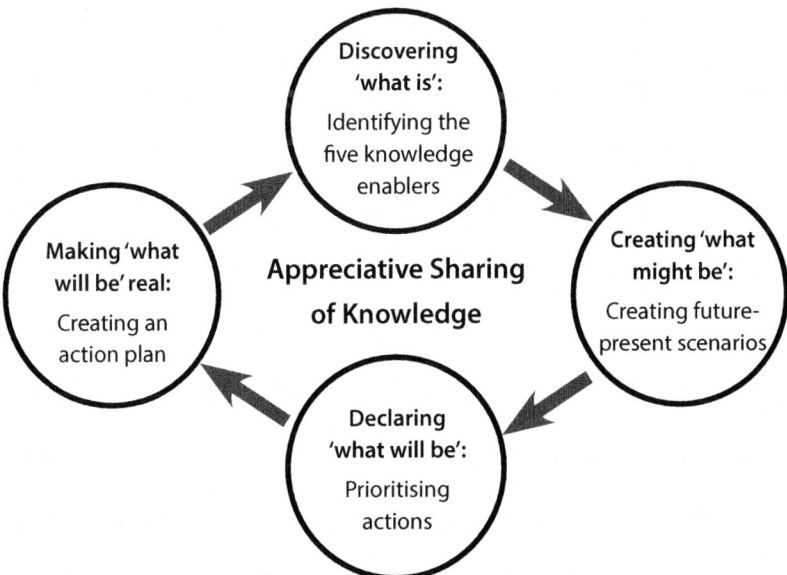

Figure 8.2: Cycle of Appreciative Sharing of Knowledge

It would be beyond the scope of this book to discuss each of the steps in a lot of detail. Only a very brief summary of what the four steps of Appreciative Sharing of Knowledge incorporate will therefore be provided here, for further details and examples, see Thatchenkery and Chowdhry (2007):

Step 1 is about 'discovering what is.' This involves getting top management or somebody else on the team who may act as a knowledge champion on board, in order for them to endorse the initiative. The Appreciative Sharing of Knowledge initiative can then be presented to the rest of the team, as well as to internal and maybe even external stakeholders. Most importantly, during this first step, an initial set of knowledge enablers will be identified through interviewing people on the team and collecting their stories. Storytelling is a very powerful tool in the process and will be discussed in more detail in the next section. Questions at this stage should be very positively worded, for example, "Think about a time when you shared something that you knew, which enabled you and your company to achieve success. Describe one such event when you felt most alive, excited, valued, or appreciated" (Thatchenkery & Chowdhry, 2007: 55). As participants share their stories of success, knowledge enablers can be identified, such as collegiality, teamwork, valuing autonomy, opportunities for personal growth, empowerment, respect, trust, or ethical behaviour (doing the right thing), to mention a few. Ideally, the facilitator will then narrow them down to four or five key knowledge enablers that are particularly relevant to the organisation and are an important part of the organisational culture.

Step 2 of the Appreciative Sharing of Knowledge approach is about 'creating what might be'. Here, further interviews will be conducted, ideally one-to-one with people who were not involved in the first phase. The identified knowledge enablers can now be explored more in-depth through stories about occasions or events where they were experienced, and through questions about factors or conditions that facilitated or promoted these knowledge enablers. Through analysing this kind of data, the knowledge infrastructure – the backbone of any knowledge enabler – can be identified, such as, for example; decision-making, organisational practices and routines, incentives for knowledge sharing, leadership, and communication. The knowledge infrastructure and the knowledge enablers are slightly different, but they are, of course, also interrelated. From this, a matrix structure can be created and specific future-present scenarios developed. These are statements based on the best of 'what is' and they envision what might be possible in the future. They should be creative, bold, and inspiring, rich and detailed, rather than restating what is already known. For example, a key knowledge enabler within an event organisation could be 'trust among the team'. When combined with the knowledge infrastructure element of 'decision-making', a basic future-present scenario could be identified as follows: 'We trust all permanent, seasonal members and volunteers of our event organisation to make the right decisions with the organisation's best interest at heart at all times.' This is a bold statement and can be adapted or made even more specific if need be, or indeed, some creative elements can be added to it to make it even more inspiring.

Step 3 in the process is about 'declaring what will be' and prioritising certain actions that have come out of the above mentioned future-present scenarios. These should be written down and shared with everyone. Lastly, step 4 involves 'making what will be real' through the creation and application of a specific action plan for all stakeholders.

It is important to note once again, that the entire Appreciative Sharing of Knowledge process is based on the current culture and climate of the organisation and based on what already works well. It does not necessarily start or stop at a specific time, in some cases it might even be worth going through the process more than once. In event organisations, for example, going through it during different times across the event cycle (pre/during/post) might be beneficial. Naturally more or less people will be involved during these different stages. Events are constantly changing, hence an approach to knowledge management that takes this ever-changing context and complex environment into account, can be very beneficial.

Using stories and storytelling to share knowledge with others

With people at the centre of knowledge sharing activities, storytelling has become a particular focus for many knowledge management researchers as well as practitioners. Through storytelling, concepts, principles, beliefs, traditions and many other things can be shared with others, both inside as well as outside the organisation. Storytelling can be a formal process (e.g. as part of a meeting, where everybody is asked to share their latest success story), but in most cases, storytelling happens informally, for example, during a coffee break, over lunch, or in the corridor (Stadler & Fullagar, 2016). Stories are low cost, usually easy to understand and share, they can be made interesting and therefore memorable (Wiig, 2004). They are also crucially important in sharing a common understanding of 'how things are done' in a specific organisation, or as Wiig (2004: 114) put it, "[s]ome reference models that people share are culturally embedded as stories or conventions that describe 'this is the way we do it here'."

In relation to events and festivals specifically, Katzeff and Ware (2006) for example, built a storytelling booth for festival volunteers to use anytime they wanted to. Volunteers were encouraged to go into the booth and record their feelings, concerns or even problems. All the stories were captured and the festival organisation learned that problems of communication, for example, led to volunteers not having the relevant information available to them at the right time. At the same time, other members of the organisation did not know about the volunteers' knowledge, expertise and skills in certain areas, as the volunteers never got an opportunity to share those with others. Through the use of this storytelling booth, volunteers' personal stories and experiences with the festival, their work and their roles were therefore made visible and explicit.

Storytelling in itself is considered a knowledge practice for employees to engage in, so once again, this fits nicely with the practice-based approach to knowledge management; but within the Appreciative Inquiry literature more specifically, knowledge management and knowledge sharing are also all about the use of language, but most importantly about the use of *positive* stories. A story can therefore be defined as, "an oral or written performance involving two or more people interpreting past or anticipated experience" (Boje, 1995: 1000). Some suggest a rather formal structure for storytelling, whereas others argue that stories change over time and they do not necessarily require a formal beginning, middle and end. They can, for example, simply be "anecdotes of experience" (Orr, 1996: 125), and the way they are told can be very different in different situations, sometimes with more details and context than at other times. Küpers (2005: 121) agreed that, "as a story is told and retold, it changes, and so the knowledge embodied in it is constantly being developed and built on." If employees already have a mutual understanding of most work tasks, routines and activities, some elements of the story do not necessarily need to be repeated every time in order to construct an actual 'story' with an introduction, point to be made and lesson learned. The gaps can be filled in by members of the team themselves without missing the key point of the story. For outsiders, however, the stories may at times seem cryptic without the required context (Orr, 1996).

The main aim of telling stories in an organisation is of course to share tacit knowledge, knowledge that requires some sort of context or meaning in order for others to acquire and use appropriately (Snowden, 2000; Gorelick et al., 2004; Küpers, 2005; Thatchenkery & Chowdhry, 2007). At the same time, through storytelling, the organisation's collective knowledge can also be increased (Whitney & Trosten-Bloom, 2003). Sharing stories about what works well in the organisation, what people value and what they hope for in terms of the organisation's future, is therefore an individual, group as well as organisational practice. Through an emphasis on stories of success and achievement, the storyteller usually elaborates on what is already working well within the organisation and provides and shares ideas of what else could work well in the organisation (Van Tiem & Rosenzweig, 2006). Using positive language and emotions is particularly useful here (Yoder, 2004). For example, "when we want more collaboration across boundaries, stories of successful collaborations are likely to get us there" (Finegold et al., 2002: 244).

Through sharing stories, the knowledge enablers and knowledge infrastructure as discussed in the steps of the Appreciative Sharing of Knowledge approach in the previous section, can be identified, such as for example, teamwork, participation, collaboration, empowerment, trust or sense of community (Bergeron, 2003; Thatchenkery & Chowdhry, 2007). Furthermore, important elements of organisational structures and culture that support knowledge management in organisations can also be shared, such as decision-making, organisational practices and routines, incentives for sharing knowledge, leadership and open communication (Thatchenkery & Chowdhry, 2007). Organisations can therefore benefit from creating opportunities for sharing stories as well as embedding storytelling into organisational processes and formal or informal rituals, so that sharing their stories with each other becomes an ongoing practice for all employees (Cooperrider & Whitney, 1999; Thatchenkery & Chowdhry, 2007). Within an event organisation, stories of previous event highlights, successes, or moments of pride and joy can, for example, be shared with new staff members and volunteers during induction sessions. This will help them understand the values and culture within the organisation and thus acquire some of the tacit knowledge necessary for the job. Stories can also help them understand how power relations can be negotiated within the organisation, as discussed in Chapter 7; for example, different stories can create multiple or even conflicting realities about the same event and can be interpreted in many different ways.

Informal ways of sharing stories are quite common in event organisations due to the fact that many seasonal employees and volunteers are out in the field, and there is not enough time to hold meetings with them and more formally share these stories. Stories can be disseminated in many different ways though, some more formal than others. For example, they can be printed in the organisation's newsletter, quoted in marketing or recruiting materials, displayed on posters, whiteboards or websites, or told and retold over lunch and coffee (Whitney & Trosten-Bloom, 2003). Stories of success quite often show what is really important to an event, the core values that employees share, as well as the goals of what the event organisation wants to achieve. It is important to get these stories across to all employees, of course, but they can also be a valuable tool to share with the attendees or wider public. For example, Glastonbury Festival prides itself in being 'green' and environmentally sustainable. Core values around these principles are shared across the team, but they are also frequently communicated with the general public to reinforce the message of success across a range of different channels. A tweet from July 2019, for example highlights: "Just heard that 99.3% of all tents were taken home. That is absolutely incredible... HUGE thanks to the record numbers who loved the farm and left no trace! #Glastonbury2019" A success story like this will inevitable make the event managers and team proud, remind them of what already works well at the festival, and lead the way in being even more environmentally sustainable in the future.

It is important to remember that once a story is shared with somebody else on the team, employees will engage in meaning-making. The same story might however be retold later on in a completely different way, within a context that makes sense to the new storyteller rather than to the original storyteller. The process of meaning-making is therefore ongoing; it takes place over time and focusses on different experiences within their cultural context (Cooperrider et al., 1995). It can lead to deeper levels of dialogue, the creation of new knowledge, and organisational learning in the long term (Sinclair, 2005).

In many organisations, and in event organisations in particular, storytelling is very much taken for granted and not recognised as a specific knowledge practice. In terms of knowledge management, storytelling therefore needs to be made more explicit and incorporated as a specific knowledge practice for employees to regularly engage in. If it becomes part of how employees work with each other, how they communicate, and engage in dialogue, then stories become a very useful tool in creating a collaborative organisational culture that, in turn, is important for effective knowledge management. In the long-run, storytelling can also become a practice for everyone to engage in throughout the year, not merely during the stressful lead-up to the actual event. Reflection after the event has taken place, for example, can be very beneficial in terms of knowledge management too. Anderson et al. (2008: 52) maintained that in an Appreciative Organisation reflection is "(...) a learning step; it is important to gauge the ways in which a shared vision has been fulfilled. However, it is also here that seeds are planted for the new stories or narratives that will give energy to future group discussions." Through ongoing collaboration and inquiry, employees can move from reflection on an individual level to reflection as a collaborative process which in turn provides opportunities for organisational learning (Loughran, 2010). Practising organisational learning from what works well within the organisation therefore needs to be ongoing, despite the fact that individual staff members join and leave the event organisation at different points in time. Stories need to be shared within and across the teams and together reflected upon in order for the organisation as a whole to be able to learn over time.

Study and discussion questions

☐ Debate how a prospective approach like Appreciative Sharing of Knowledge could be a better approach to knowledge management in an event organisation than retrospective, problem-solving approaches.

☐ Using an example from your own experience, work through the cycle of Appreciative Sharing of Knowledge for an event organisation.

☐ Find online a 'story' about an event of your choice and discuss how this story creates a shared sense of meaning around the event, what the event is about, and how it is talked about by others (e.g. on social media, blogs or forums).

☐ Referring back to the 'pulsating' nature of event organisations introduced in Chapter 2, provide an example of how storytelling can be used as a knowledge management tool at different stages of the event cycle.

Recommended additional readings

Boje, D. M. (1995). Stories of the storytelling organization: A Postmodern analysis of Disney as "Tamara-Land". *Academy of Management Journal*, 38(4), 997-1035.

Thatchenkery, T., & Chowdhry, D. (2007). *Appreciative Inquiry and Knowledge Management - A Social Constructionist Perspective*. Cheltenham: Edward Elgar.

Whitney, D., & Trosten-Bloom, A. (2003). *The Power of Appreciative Inquiry - A Practical Guide to Positive Change*. San Francisco: Berrett-Koehler.

References

Anderson, H., Cooperrider, D. L., Gergen, K. J., Gergen, M., McNamee, S., Watkins, J. M., & Whitney, D. (2008). *The Appreciative Organization* (revised ed.). Chagrin Falls, Ohio: Taos Institute Publications.

Bergeron, B. (2003). *Essentials of Knowledge Management*. Hoboken NJ: John Wiley & Sons, Inc.

Boje, D. M. (1995). Stories of the storytelling organization: A Postmodern analysis of Disney as "Tamara-Land". *Academy of Management Journal*, 38(4), 997-1035.

Bushe, G. R. (2007). Appreciative Inquiry is not (just) about the positive. *OD Practitioner*, 39(4), 30-35.

Cooperrider, D. L., & Srivastva, S. (1987). Appreciative Inquiry in organizational life. *Research in Organizational Change and Development*, 1(1), 129-169.

Cooperrider, D. L., & Whitney, D. (1999). *Collaborating for Change: Appreciative Inquiry*. San Francisco, CA: Barrett-Koehler.

Cooperrider, D. L., Barrett, F. T., & Srivastva, S. (1995). Social Construction and Appreciative Inquiry: A journey in organizational theory. In D. M. Hosking, H. P. Dachler, & K. J. Gergen (Eds.), *Management and Organization: Relational Alternatives to Individualism*. Ashgate Publishing.

Finegold, M. A., Holland, B. M., & Lingham, T. (2002). Appreciative Inquiry and Public Dialogue: An approach to community change. *Public Organization Review*, 2(3), 235-252.

Gorelick, C., Milton, N., & April, K. (2004). *Performance through Learning - Knowledge Management in Practice*. Amsterdam: Elsevier Butterworth-Heinemann.

Katzeff, C., & Ware, V. (2006). *Video Storytelling as Mediation of Organizational Learning*. Paper presented at the NordiCHI2006: Changing Roles, Oslo.

Koster, R. L. P., & Lemelin, R. H. (2009). Appreciative Inquiry and rural tourism: A case study from Canada. *Tourism Geographies*, 11(2), 256–269.

Küpers, W. (2005). Phenomenology of embodied implicit and narrative knowing. *Journal of Knowledge Management*, 9(6), 114-133.

Loughran, J. (2010). Reflection through collaborative action research and inquiry. In N. Lyons (Ed.), *Handbook of Reflection and Reflective Inquiry - Mapping a Way of Knowing for Professional Reflective Inquiry* (pp. 399-413). New York: Springer.

Maier, T. A. (2008). Appreciative Inquiry and hospitality leadership. *Journal of Human Resources in Hospitality & Tourism*, 8(1), 106–117.

Orr, J. E. (1996). *Talking about Machines - An Ethnography of a Modern Job*. Itaca: Cornell University Press.

Raymond, E. M., & Hall, M. C. (2008). The potential for Appreciative Inquiry in tourism research. *Current Issues in Tourism*, 11(3), 281-292.

Rogers, P., & Fraser, D. (2003). Appreciating Appreciative Inquiry. *New Directions for Evaluation*, 100, 75-83.

Sinclair, J. (2005). The impact of stories. *Electronic Journal of Knowledge Management*, 3(1), 53-64.

Snowden, D. (2000). New wine in old wineskins: from organic to complex knowledge management through the use of story. *Emergence*, 2(4), 50-64.

Stadler, R., & Fullagar, S. (2016). Appreciating formal and informal knowledge transfer practices within creative festival organizations. *Journal of Knowledge Management*, 20(1), 146-161.

Thatchenkery, T., & Chowdhry, D. (2007). *Appreciative Inquiry and Knowledge Management - A Social Constructionist Perspective*. Cheltenham: Edward Elgar.

Van der Haar, D., & Hosking, D. M. (2004). Evaluating Appreciative Inquiry: A relational constructionist perspective. *Human Relations*, 57(8), 1017-1036.

Van Niekerk, M., & Getz, D. (2019). *Event Stakeholders*. Oxford: Goodfellow Publishers.

Van Tiem, D., & Rosenzweig, J. (2006). *Appreciative Inquiry*. ASTD Press.

Whitney, D., & Trosten-Bloom, A. (2003). *The Power of Appreciative Inquiry - A Practical Guide to Positive Change*. San Francisco: Berrett-Koehler.

Wiig, K. M. (2004). *People-Focused Knowledge Management – How Effective Decision Making Leads to Corporate Success*. Amsterdam: Elsevier Butterworth Heinemann.

Yoder, D. M. (2004). Organizational climate and emotional intelligence: An Appreciative Inquiry into a 'leaderful' community college. *Community College Journal of Research and Practice*, 29(1), 45-62.

9 Practical Implications and Recommendations for Event Organisers

Learning objectives

- Explore recommendations for event managers to implement effective knowledge management strategies within their organisations.
- Understand appropriate knowledge management activities and practices for each stage of the event management process (pre/during/post event).
- Identify areas for future research into knowledge management in event organisations.

Introduction

Throughout this book a number of practical implications and recommendations for event organisers have been mentioned and outlined. This chapter aims to bring them all together in relation to some of the challenges faced by event organisations specifically, as covered in Chapter 2. It is worth remembering though, that any knowledge management initiative in any kind of organisation is not just down to top management and their aims and objectives. Debowski (2006: 337) nicely summarised the four key knowledge management principles as follows:

- Knowledge management is everyone's business;
- Knowledge practices are legitimate core business;
- Communication is essential; and
- High performance should be encouraged.

While the recommendations presented throughout this chapter are largely aimed at event managers, they need to be incorporated with all employees in mind; permanent and seasonal staff members, volunteers, as well as contractors, suppliers, local businesses and other stakeholders. An event organisation can only learn as a whole over time by effectively bringing together knowledge and expertise from as many different sources as possible. Furthermore, effective knowledge management, and a more explicit understanding of it, will enhance professionalisation across the events industry and will make event professionals more employable in the future (Stadler et al., 2014).

The book has also highlighted that there is still a need for further research into knowledge management in event organisations. In the final section of this chapter, suggestions for future research will hence be presented. This could be in the form of student research projects, dissertations (undergraduate or postgraduate), as well as consultancy work or other types of applied research.

Knowledge management recommendations for event organisers

The 'pulsating' nature of events requires event managers to think about the different stages of their event and hence different knowledge management strategies, processes and practices during each stage (Clayton, 2020). These recommendations are based on the kind of knowledge practices can be implemented pre-event, during the event and post-event, in order to maximise organisational learning in the long run. It is important to remember that these strategies sometimes overlap and do not necessarily start and stop during a specific phase. In fact, it is best to think of some knowledge practices to be implemented pre-event, but then to continue throughout the later stages, perhaps reinforcing them or adapting them as and when necessary. Similarly, the knowledge management activities staff members engage in during the event will be beneficial to maintain post-event and perhaps even carry over into the next event season, where in turn, pre-event knowledge management can then start at an already higher and more complex level than the year before. Figure 9.1 below illustrates the recommended knowledge management activities and practices throughout the different stages.

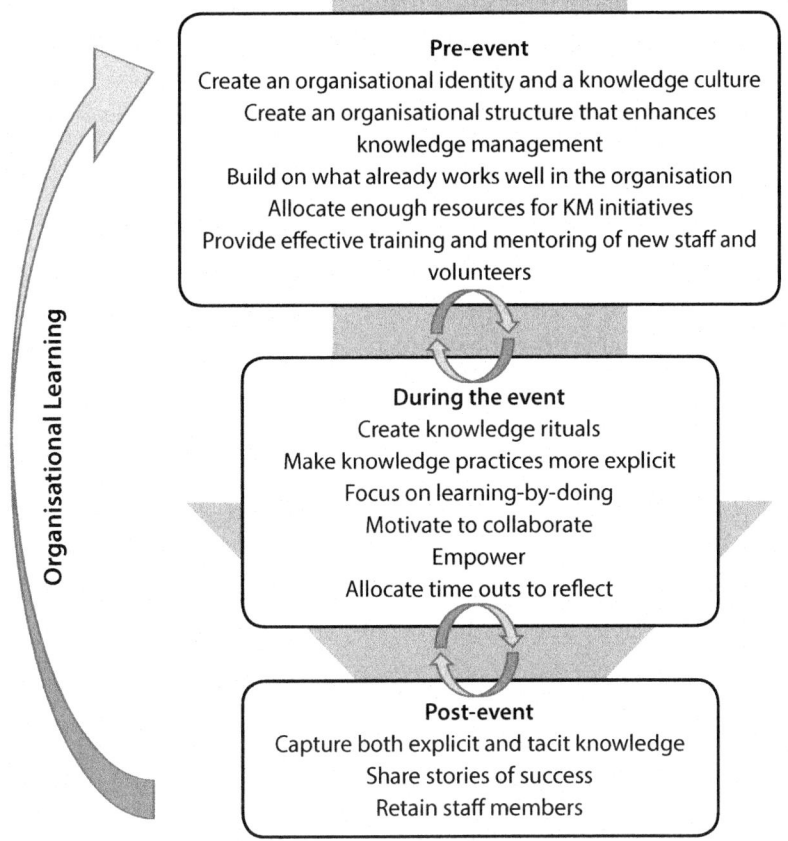

Figure 9.1: Knowledge management recommendations for event managers

Pre-event knowledge management and knowledge practices

As highlighted throughout this book, any knowledge management initiative needs to be built into a broader strategy for the organisation as a whole. Ideally, knowledge management efforts need to be aligned with the existing culture and structure of the organisation, otherwise they will be of limited success (Du Plessis, 2006; Chen & Huang, 2007; Intezari et al., 2017). In the months leading up to an event and with only the permanent team around, it is therefore crucial to create and reinforce an organisational vision, mission and identity, which help staff members identify themselves with the organisation (*I am a member of this organisation and it is important to me*). Equally important is the creation of an organisational culture (*this is who we are and this is the way things are done in our organisation*), including beliefs, behaviours and values that are shared by all organisational members; and lastly, an organisational climate (*the way we interact and work with each other*) that enhances knowledge management activities and practices. Some ideas of how this could be expressed and of how a shared meaning can be created, were presented as part of the QMF Case Studies. An open and collaborative culture and climate based on shared values, beliefs and ideals are the basis for effective knowledge management, and if it does not exist amongst the core and permanent team, it will most likely not be shared with the seasonal staff and volunteers, suppliers, contractors and other stakeholders either.

At the same time, it is crucial to have an organisational structure in place that supports knowledge management activities, processes and practices, and is ready to expand and contract as and when new staff members join the team. Again, it is beneficial to think about this before seasonal staff and volunteers come on board. The structure should ideally emphasise teamwork and collaboration, rather than top-down, hierarchical approaches to management, which can stop the flow of knowledge between different hierarchical levels. To think of key middle managers as the link between directors and permanent staff on the one hand, and seasonal staff and volunteers on the other hand, is a good idea here, and is in line with Nonaka and Takeuchi's (1995) suggested middle-up-down approach to knowledge management. Interdisciplinary pods or the development of communities-of-practice should also be actively encouraged, in order to break down silos and to enhance knowledge flow between teams and departments. Perhaps it is even possible to identify and assign specific knowledge management roles and responsibilities to certain staff members, such as the ones suggested in Chapter 5. Having a chief knowledge officer and several knowledge brokers on the team, for example, can help make knowledge management more explicit and bring it to employees' attention early on.

It is important to remember that to completely start from scratch with all of these strategies, is not feasible and definitely not desirable. Any event organisation aiming to strengthen their knowledge activities and practices, should start with identifying what already works well within the organisation and then build on that. In events particularly, stories of success from previous event experiences can be exciting to share and relive with the team, before embarking on the next event journey. The Appreciative Sharing of Knowledge approach can be a useful tool here to identify internal knowledge management strengths, as well as the resources, activities and messages necessary to develop them further (Thatchenkery & Chowdhry, 2007).

Whilst it should be acknowledged that knowledge management initiatives can be resource- and time-intensive – factors that pose a significant challenge for events, and in particular small-scale, or volunteer-run events – setting aside some resources for knowledge management early on, will pay off in the long run. It is crucial for the senior management team to think through all stages of the event planning and delivery process, so as to not run out of resources post-event when ideally knowledge will need to be captured and stored for future years, as discussed further below.

Lastly, enough resources also need to be allocated for training and mentoring new staff and volunteers as and when they join. It has been highlighted in this book that knowledge management and HR management need to be closely aligned and any efforts to train new staff members should be run with knowledge management activities and practices in mind (Gloet & Berrell, 2003; Clayton, 2016, 2020; Lockstone-Binney et al., 2020). Rather than, for example, bringing in volunteers at the last minute or even on the day of the event, and going through a very brief induction session barely covering the basics with them, it is more beneficial to engage them early on in the process, identify their strengths, skills, and capabilities, as well as their needs and wants, which will ensure they stay motivated and they will create new knowledge together, share it with others, and use it effectively. Setting up a buddy- or mentor-system might also work in some event organisations. Clayton (2016) for example, highlights the importance of having a few 'old-timers' available – more experienced volunteers who have done similar jobs before – to mentor and teach new volunteers (similar to a master-apprentice type of relationship). This more informal process of sharing knowledge helps develop special bonds and a sense of community among volunteers and is an important 'soft' factor for knowledge management.

Hanlon and Cuskelly (2002) recommend that an induction kit can further help facilitate this induction and training process. It should include items such as: an annual report, a welcome message from the

CEO, a name badge, a staff list, a uniform (t-shirt), a list of stakeholders and locations, and a detailed induction manual with an overview of the event, maps, policies, schedules, rotas, and contact lists. In other words, volunteers need to have a good understanding of key aspects of the event, who is who, where to find relevant information, and what to do in an emergency. These kinds of questions can be provided in documented form and thus constitute some of the explicit knowledge that needs to be shared with volunteers. It is, however, difficult to assess whether volunteers will actually know how to use this information and how to apply it to their roles. Teaching them the more tacit aspects of getting the job done is, again, sadly often overlooked.

According to Clayton (2016), in addition to these manuals and training sessions, good leadership and supervisory support can help with making volunteers feel more engaged, prepared and confident. Clear communication, a culture of collaboration and a sense of community also enhance this experience for volunteers and help them understand and learn. Blackman et al. (2017) similarly argued that volunteers need good listening and communication skills based on respect for each other in order to be able to effectively share knowledge. It is important to allocate time and resources for volunteers to acquire these skills pre-event, so they can further build on them during and post-event. Rather than merely bringing volunteers in to 'get the job done', an opportunity for them to develop new knowledge and skills can be an intrinsically motivating and very rewarding factor, which ideally leads to continuance commitment, where they will then be more inclined to return the following year and can be better matched with roles and tasks they already know how to perform (Clayton, 2016).

Knowledge management and knowledge practices during the event

Once the event is in full swing, it becomes more and more difficult to allocate time for knowledge management activities, and especially knowledge transfer often becomes an ad-hoc activity (Muskat & Deery, 2017). With the above elements already in place, it is, however, a good idea to now create and encourage a range of knowledge rituals both within teams and groups, and across different teams, whenever possible. This can range from formal meetings (e.g. a brief team meeting in the morning, then another one in the afternoon), to any ritual-like activity such as informal catch-ups over lunch or coffee, or checking in with each other in the corridor or even backstage (Smith & Stewart, 2011; Stadler & Fullagar, 2016). Rituals can reinforce a collaborative culture based on trust, teamwork, and motivation, and they are key in sharing both explicit and tacit knowledge (Armistead & Meakins, 2002; Ragsdell & Jepson, 2014).

Furthermore, while checklists, to-do lists, manuals, and databases are of course important tools for information and knowledge management, this book has highlighted that technology can never be the answer to everything. Alongside these documented forms of knowledge, it is therefore also important to make knowledge practices – and especially knowledge sharing practices during the event itself – more explicit. This will help staff members, volunteers and other stakeholders understand how things are done, acquire the relevant know-how as well as the embedded and embodied type of knowledge. They need to be aware of the relational and practice-based understanding of knowledge management, where knowledge is produced, enacted, embodied and shared (Orlikowski, 2002; Hislop et al., 2018). A particular emphasis should thereby be put on learning-by-doing approaches to knowledge management and knowledge sharing. This will help new staff members immerse themselves in the practice itself, as well as enhance their social interaction with others, such as with a mentor, for example.

During the event itself it is also impossible to achieve tasks by working alone. Any opportunity to collaborate should therefore be encouraged, as it reduces the risk for knowledge hiding and hoarding to become the norm. Praising and highlighting positive examples of teamwork can be intrinsically motivating for staff members to work together even more and to effectively share knowledge with each other (Hung et al., 2011). The sum should be greater than the parts here; the team's knowledge overall will be more than what each individual team member can contribute to the task. Senior managers and permanent staff members can make a conscious effort to encourage this kind of thinking through effective leadership, leading by example and practising similar behaviour within their own teams. It can also be rewarding for staff members and volunteers to be mentioned in newsletters, success stories shared during meetings, or on social media. By including positive examples of knowledge practices in these stories, other team members are then likely to also make a more conscious effort to engage in knowledge management activities and practices.

Senior management and team leaders can furthermore benefit from thinking about power as a positive resource that can be used to engage staff members in knowledge activities. It has been shown in Chapter 7 that it is important to acknowledge not only the understanding of 'knowledge is power', but to also look into 'power is knowledge' (Gordon & Grant, 2005). This requires senior management to give up some of their power, which they might feel reluctant to do. Creating a less hierarchical structure in the first place, as discussed earlier, can help with this. Using power in a positive sense and empowering others within the organisation can then further create a sense of commitment and ownership, responsibility over decisions made, and hence more knowledge about the job itself as well as its wider implications for the organisation as a whole. By redistributing power among all members of the organisation, both vertical as well as horizontal knowledge practices can be enhanced (Mosley et al., 2001). Other stakeholders should also be empowered, especially local or marginalised communities in order to make their voices heard and for them to truly own the event. This can, in turn, create new knowledge for the event organisation.

Last but not least, during the event there is little time to sit back and reflect on knowledge management activities, processes and practices. The stressful and hectic environment often means that staff members and volunteers simply jump in and help out whichever way possible, but this might not always be the most efficient or effective way of accomplishing a task. It is suggested that team leaders, staff members and volunteers take regular 'time outs', not only to recharge their batteries, but also to ponder which knowledge practices work well within the organisation and which ones might need to be adapted. This will help the organisation learn over time.

Post-event knowledge management and knowledge practices

Post-event, the main emphasis should be on evaluating the overall performance of the event, and how successfully it has achieved its goals. This includes, "information gathering and feedback through which processes can be improved, goals more effectively attained, and by which organisations can learn and adapt" (Getz, 2018: 2). In relation to knowledge management, the focus should therefore be on capturing both explicit and tacit knowledge and building an organisational memory before staff members and volunteers leave and hence take the newly created knowledge with them. However, at this point there is usually little time and few resources left to write everything down and store it for years to come (Muskat & Deery, 2017). An effective way of capturing some of this knowledge is through debriefing events or exit interviews. These can be done formally with a team leader or senior manager, or they can be recorded by staff members and volunteers themselves in a more informal way (e.g. in a digital diary), then collected, transcribed and analysed in order to identify challenges, issues, and ideas for improvement. If possible, staff members could even regularly record their thoughts and feelings during the event itself (e.g. on the bus journey to/from work), then post-event they could share their highlights with the team. It requires commitment from everyone, but could be turned into a key exercise everybody will engage in, if there is, for example, a prize or reward for the best story and most innovative idea of how things could be improved in the future.

As outlined in Chapter 8, focusing on stories of success is always a good idea as it engages and motivates employees, and helps identify ways of being even more successful (Cooperrider & Whitney, 1999; Thatchenkery & Chowdhry, 2007). Challenges and issues also need to be addressed, of course, but should ideally be turned into opportunities for learning. Many staff members are, at this stage, very proud of the event experience they have helped create, and will want to share this sense of achievement, pride and success with others. Asking specific questions around what kind of knowledge they needed to acquire in order for this to happen, how they used it, and perhaps how it could be stored for the future, will once again make knowledge management more explicit and will make everybody aware of its importance in the long-term success of the organisation as a whole. This can also create a human capital legacy, as suggested by Blackman et al. (2017).

Lastly, creating opportunities for recognition and reward can also help retain staff members and volunteers, or encourage them to come back the following year. It creates a sense of organisational loyalty and identity, and ensures that knowledge will not leak into other event organisations. Keeping them engaged throughout the year can also help in creating this sense of loyalty, while at the same time the organisation as a whole can benefit from tapping into stories of knowledge management successes which these seasonal staff members still remember, but may have been forgotten by the permanent team. Regular meetings or social events with them throughout the year are recommended.

Future research into knowledge management in event organisations

Throughout this book it has been highlighted that there has been a lot of research on knowledge management in general and across different types of companies and organisations. Yet, there is still a lack of research into knowledge management within event organisations and hence plenty of opportunities to explore these topics further through research projects, such as student dissertations, collaborative research across different institutions and/or with industry partners, or consultancy work. Table 9.1 below (based on Stadler, 2019) presents areas that have already been explored in detail in the events literature, as well as gaps in research and specific ideas for future research:

Table 9.1: Past and future research into knowledge management in event organisations

Topics already covered in the events literature	Opportunities for future research
Knowledge management in mega-sport events and music festivals	Knowledge management in other types of events, such as conferences, conventions, product launch events, or virtual events
Knowledge sharing/transfer among volunteers	Other knowledge management activities volunteers engage in (e.g. knowledge creation or storage)
Knowledge sharing/transfer among permanent and paid staff	The role of other stakeholders (both primary and secondary) in knowledge sharing/transfer practices, to gain a more holistic view on knowledge management and appropriate roles and responsibilities
Knowledge management in relation to operational practices	Knowledge management in relation to creative practices, as well as research on the similarities and differences around creative/artistic knowledge practices vs. operational/ strategic knowledge practices

Knowledge documentation and storage, using technology and databases	More research focusing on the relational and practice-based understanding of knowledge management, the 'knowing' and 'know how'
Knowledge (and power) as static concepts	Knowledge (and power) as dynamic, and therefore as both positive and negative resources in event organisations
Short-term studies, mostly during and post-event	Longitudinal studies covering pre, during and post-event stages, and more than just one full event cycle
Use of qualitative research methods (interviews, focus groups, diaries, participant observation, and/or document analysis) within a single-case study approach	Studies using mixed methods and multidimensional research approaches, both within single case studies as well as multiple/comparative cases across different geographic areas, different event sizes/types, or different organisational structures

Specific research questions can be developed from within one such area, or through a combination of two or more of them. For example, the difference between creative/artistic knowledge practices and operational knowledge practices could be investigated within a specific type of event, such as a product launch event or a virtual event. Another example would be to explore the role of suppliers as a key stakeholder group with regards to knowledge management in event organisations. A focus could thereby be on specific knowledge management activities they engage in, such as the acquisition of knowledge, the creation of new knowledge, or the transfer of that knowledge onto the event organisation and/or other stakeholders. This can be investigated using a range of (qualitative and/or quantitative) methods, across several case studies (perhaps several different event organisations that these suppliers work with), as well as over time.

It is clear to see that knowledge management is an exciting topic for further research. Any future research will help contribute to our understanding of the complex knowledge processes and practices found within a very challenging event environment, with little knowledge management resources available, and little time to reflect. Making knowledge management a more explicit part of event management will enhance organisational learning and hence provide a competitive advantage to event organisations in the long term.

Study and discussion questions

☐ Put together a knowledge management checklist or a spreadsheet for event managers to use pre-, during and post-event.

☐ Provide a list of knowledge management Dos and Don'ts for new staff members and volunteers in an event organisation. What are your top tips for them?

☐ Identify three specific research questions for future research into knowledge management in event organisations.

Recommended additional readings

Muskat, B., & Deery, M. (2017). Knowledge transfer and organizational memory: an events perspective. *Event Management*, 21(4), 431-447.

Stadler, R. (2019). Knowledge management in event and festival organisations: challenges and future directions. In E. Lundberg, J. Armbrecht, & T. Andersson (Eds.), *A Research Agenda for Event Management* (pp. 154-169). Edward Elgar.

References

Armistead, C., & Meakins, M. (2002). A framework for practising knowledge management. *Long Range Planning*, 35(1), 49-71.

Blackman, D., Benson, A. M., & Dickson, T. J. (2017). Enabling event volunteer legacies: a knowledge management perspective. *Event Management*, 21(3), 233-250.

Chen, C.-J., & Huang, J.-W. (2007). How organizational climate and structure affect knowledge management - The social interaction perspective. *International Journal of Information Management*, 27(2), 104-118.

Clayton, D. (2016). Volunteers' knowledge activities at UK music festivals: a hermeneutic-phenomenological exploration of individuals' experiences. *Journal of Knowledge Management*, 20(1), 162-180.

Clayton, D. (2020). Knowledge management in events. In S. J. Page & J. Connell (Eds.), *The Routledge Handbook of Events* (2nd ed., pp. 442-456). London: Routledge.

Cooperrider, D. L., & Whitney, D. (1999). Collaborating for Change: Appreciative Inquiry. San Francisco, CA: Barrett-Koehler.

Debowski, S. (2006). *Knowledge Management*. Australia: John Wiley & Sons Ltd.

Du Plessis, M. (2006). *The Impact of Organisational Culture on Knowledge Management*. Oxford: Chandos Publishing.

Getz, D. (2018). *Event Evaluation*. Oxford: Goodfellow Publishers.

Gloet, M., & Berrell, M. (2003). The dual paradigm nature of knowledge management: implications for achieving quality outcomes in human resource management. *Journal of Knowledge Management, 7*(1), 78-89.

Gordon, R., & Grant, D. (2005). Knowledge management or management of knowledge? Why people interested in knowledge management need to consider Foucault and the construct of power. *Tamara: Journal of Critical Postmodern Organization Science, 3*(2), 27-38.

Hanlon, C., & Cuskelly, G. (2002). Pulsating major sport event organizations: A framework for inducting managerial personnel. *Event Management, 7*(4), 231-243.

Hislop, D., Bosua, R., & Helms, R. (2018). *Knowledge Management in Organizations - A Critical Introduction* (4th ed.). Oxford: Oxford University Press.

Hung, S.-Y., Durcikova, A., Lai, H.-M., & Lin, W.-M. (2011). The influence of intrinsic and extrinsic motivation on individuals' knowledge sharing behavior. *International Journal of Human-Computer Studies, 69*(6), 415-427.

Intezari, A., Taskin, N., & Pauleen, D. J. (2017). Looking beyond knowledge sharing: An integrative approach to knowledge management culture. *Journal of Knowledge Management, 21*(2), 492-515.

Lockstone-Binney, L., Hanlon, C., & Jago, L. (2020). Staffing for successful events: Having the right skills in the right place at the right time. In S. J. Page & J. Connell (Eds.), *The Routledge Handbook of Events* (2nd ed., pp. 427-441). London: Routledge.

Mosley, D. C., Megginson, L. C., & Pietri, P. H. (2001). *Supervisory Management - The Art of Empowering and Developing People* (5th ed.). South Western: Thomson Learning.

Muskat, B., & Deery, M. (2017). Knowledge transfer and organizational memory: an events perspective. *Event Management, 21*(4), 431-447.

Nonaka, I., & Takeuchi, H. (1995). *The Knowledge Creating Company – How Japanese Companies Create the Dynamics of Innovation*. New York: Oxford University Press.

Orlikowski, W. J. (2002). Knowing in practice: Enacting a collective capability in distributed organizing. *Organization Science, 13*(3), 249-273.

Ragsdell, G., & Jepson, A. S. (2014). Knowledge sharing: insights from Campaign for Real Ale (CAMRA) festival volunteers. *International Journal of Event and Festival Management, 5*(3), 279-296.

Smith, A. C. T., & Stewart, B. (2011). Organizational rituals: Features, functions and mechanisms. *International Journal of Management Reviews*, 13(2), 113-133.

Stadler, R. (2019). Knowledge management in event and festival organisations: Challenges and future directions. In E. Lundberg, J. Armbrecht, & T. Andersson (Eds.), *A Research Agenda for Event Management* (pp. 154-169). Cheltenham and Northampton: Edward Elgar.

Stadler, R., & Fullagar, S. (2016). Appreciating formal and informal knowledge transfer practices within creative festival organizations. *Journal of Knowledge Management*, 20(1), 146-161.

Stadler, R., Fullagar, S., & Reid, S. (2014). The professionalization of festival organizations: A relational approach to knowledge management. *Event Management*, 18(1), 39-52.

Thatchenkery, T., & Chowdhry, D. (2007). *Appreciative Inquiry and Knowledge Management - A Social Constructionist Perspective*. Cheltenham: Edward Elgar.

Glossary

Appreciative Inquiry: A strengths-based, positive approach to leadership development and organisational change to identify 'what is', 'what might be', 'what could be' and finally 'what will be'.

Appreciative Sharing of Knowledge: A prospective approach to knowledge management that regards knowledge sharing as an opportunity to embrace and aspire to, centred around the question, 'what makes people share knowledge?'

Chief knowledge officers: (Usually) top managers, responsible for managing the entire organisational knowledge management strategy and processes at the corporate level; they identify strategies, design systems and facilitate the transfer of knowledge across the organisation.

Community-of-practice: A group of people who have a particular activity in common, and as a consequence have some common knowledge, overlapping values, and a shared identity, which allow the formation of a shared understanding of values and assumptions within the community-of-practice, as well as create social conditions which enhance knowledge sharing, creation and utilisation.

Data: (Raw) numbers without any particular meaning; discreet, objective facts.

Empowerment: The granting of authority to employees to make key decisions within their area of responsibility based on their knowledge and expertise.

Explicit knowledge: Knowledge that can be expressed in words and numbers, and can be shared and communicated easily through, e.g. formulas, principles and procedures.

Group knowledge: Knowledge that has been created and shared by individual members of a group around certain work processes and practices to develop a shared understanding among members; ideally more than the sum of all group members' individual knowledge.

Individual knowledge: Knowledge acquired over time and in different contexts, possessed by an individual, and based on their individual background, expertise and experience.

Information: Data that has been put into a certain context or arranged in a specific order; categorised, calculated, corrected or condensed data.

Interdisciplinary pods: Formal or informal work teams bringing together expertise between colleagues from different backgrounds to enable the development of new ideas and novel ways of working.

Knowledge: Data and contextualised information combined with experience, values and expert insights that is being put into practice through processes, procedures, documents or repositories, to add value to the resulting activity of an individual, team or organisation.

Knowledge acquisition: The process of sourcing knowledge that is relevant to a certain problem faced, as well as collecting and gathering knowledge and importing it into other business processes; can be formal or informal, e.g. learning-by-doing.

Knowledge brokers: Middle managers who serve as a bridge between the visionary ideals of the top and the day-to-day operational tasks of front-line workers; responsible for creating connections between different people (and between different levels) of the organisation in order for them to be able to share knowledge, as well as to move knowledge around between them.

Knowledge creation: The production and development of new ideas, the generation of new knowledge for specific tasks, in order for the organisation to evolve and innovate.

Knowledge hiding: Intentionally concealing and not sharing knowledge with others; keeping knowledge that has specifically been requested a secret from the person who requested it.

Knowledge hoarding: Unintentionally retaining or accumulating knowledge, which may or may not be shared with others at a later stage without realising that it would be of value to them.

Knowledge identification: The process of proactively identifying internal organisational knowledge; the process whereby organisations take steps to identify the relevant and needed knowledge that exists within their boundaries (e.g. Knowledge Audit or Knowledge Sourcing) through locating, finding and discovering knowledge; mapping, organising and classifying knowledge; as well as structuring, analysing and reviewing existing knowledge within the organisation.

Knowledge leakage: Competitive knowledge that intentionally or unintentionally leaves the organisation, and moves into another organisation where it can be used or applied in either the same way or in new competitive ways.

Knowledge management (in event organisations): The effective use of organisational systems, processes and practices which allow both explicit and tacit knowledge to be created, identified, acquired, utilised, shared and stored, in order for the organisation to produce a successful event experience and to sustain a competitive advantage over time.

Knowledge practice theory: 'Knowing' and 'know-how' as a social practice that acknowledges both the historical and structural context in which actions take place, where knowing and know-how are enacted through and embedded in people's everyday activities; knowledge that is embedded in work practices, tasks and routines, and embodied by the people who carry out and engage in these practices.

Knowledge practitioners/knowledge workers: Front-line workers, who use knowledge to produce value for the organisation; they create, share, embody, accumulate, and use knowledge, and they engage in a range of knowledge practices as part of their job.

Knowledge storage: The process of archiving and preserving knowledge; codifying knowledge in order to make it explicit; as well as packaging, securing and protecting knowledge to create an organisational memory and to avoid knowledge leakage.

Knowledge transfer/sharing: The process of communicating knowledge with others inside and outside the organisation; disseminating, diffusing and distributing knowledge to those who need it; as well as networking, collaborating and cooperating with others.

Knowledge use: The process of applying knowledge, exploiting and capitalising on the knowledge created and identified, as well as re-using knowledge in different situations and at different times.

Organisational culture: Beliefs and behaviours shared by an organisation's members, including ideas, values, and certain acceptable behaviours; the 'way things are done' within the organisation.

Organisational identity: A commonly held representation of the organisation amongst its members; the essence of the organisation that distinguishes it from others; 'who we are' as an organisation.

Organisational knowledge: Knowledge about the organisation's history, vision, values and beliefs, which help individuals and groups understand how they can contribute to the organisation's overall success; a shared understanding between all employees of 'who they are' and 'how things are done' within the organisation.

Organisational learning: A continuous process for the organisation as a whole to go through in order to constantly improve and innovate through the knowledge contributions of its individual and group/team members.

Organisational memory: Archived experiences, routines, processes, actions and decisions held by the organisation as a whole.

Organisational vision: A shared meaning of what the organisation is, and a mental image of the ideal state that the organisation wishes to achieve; it is inspirational and aspirational for employees and stakeholders.

Power: The capacity or ability to direct or influence the behaviour of others; a resource used by one actor 'over' another; or a complex and dynamic network of social relations, where power constantly changes and circulates within and between them. In organisations: the rules of the game within an organisation, which can both enable and constrain action.

Power/knowledge: Foucault's concept where power and knowledge are inseparable: power is based on knowledge and makes use of knowledge, but power also produces knowledge by shaping and reshaping it.

Pulsating organisations: Organisations that expand and contract, such as event organisations; they operate with a small core of personnel for much of the year, expand substantially in the lead up to an event, and then contract again once the event is over.

Ritual: Practices that become more and more routinized within an organisation, usually with symbolic meaning; can be formal (e.g. meetings), or informal (e.g. having a coffee with a colleague).

Stakeholder power: The ability of a stakeholder to use resources to make an event happen or to secure a desired outcome.

Tacit knowledge: Knowledge that is individual, relational and context-specific, part of people's actions, experiences and beliefs; taken for granted ways of doing things, which are difficult to capture and share with others.

Index

Appreciative Inquiry 156–160
  at organisation, group or team level 159
  four steps of 158
  key principles 157
Appreciative Sharing of Knowledge 155–170
  four steps of 162

bottom-up management 100

capital
  human 40
  organisational 40
  social 40
chief knowledge officers 106
chronos-time and kairos-time 23
co-creation 127–128
collaboration 127–128, 177
Colorado Music Festival 88–90
  communities-of-practice 88–90
  levels of participation 88
communities-of-practice 79–85, 174
  and formal work groups 80
  common purpose 84
  Colorado Music Festival 88–90
  critiques 86–87
  identification with 115
  levels of participation 81–83
  power issues 139
  situated learning 83
  support innovation 85
competitive advantage 1, 39, 42, 181
conflict in organisations 136–137
creative knowledge 27–28

creativity 85
culture. *See* organisational culture

data, defined 2

emotions 78–79
empowerment 144–146, 178
  links to knowledge management 145
  psychological 145–146
  Queensland Music Festival 147–151
  social-structural 144–145
event planning process 175
expertise
  lack of 23–24
  sharing of 24
expert power 137
explicit knowledge 5–6

'family' metaphor 120
forgetting 29
four knowledge flow dimensions
  festival example 62–63
functional teams
  limited knowledge transfer 103

group knowledge 6–7

HR. *See* human resource management
human capital 40
human resource management
  closely related to knowledge management 95–96
  strategies 94–112

ICT
  knowledge acquisition 46
  knowledge creation 43–44
  knowledge identification 44
  knowledge sharing/transfer 50
  knowledge storage 52
  knowledge use 48
identity, concept 114
  organisational 114–116, 173
individual knowledge 6
induction kit 175
induction sessions 18, 46
information, defined 3
innovation 85–86
  interdisciplinary teams 103
  knowledge creation 54
inter-disciplinary pods/teams
    103–104, 174
  knowledge sharing/transfer 51, 103
  Queensland Music Festival
    104–105

know-how and know-what 72
knowing, defined 72
knowing and know-how 72–73
knowledge
  acquisition 45–46
    ask an expert 45
    learning by being told 46
  brokers 106
  capture, post-event 179
  champions and strategists 106
  claims 64
  conversion 56–57
  creation 42–44
    importance of middle managers
      101
    spiral 54–57
  creative, operational and strategic
    27–28
  culture enablers 119
  defined 3

  embedded and embodied 72
  enacted 72–73
  engineers 106
  explicit 5–6
  flow functions 58–60
  hiding 31
  hoarding 31
  identification 44–45
    individual, group and
      organisational 6–8
  leakage 30
    and staff retention 99
  management. *See* knowledge
    management
  modes 54–55
  officers 106
    See chief knowledge officers
  operators 107
  practice rituals 74–75
  practitioners 107
  sharing. *See* knowledge sharing
  specialists 107
  storage 51–53
  tacit 5–6
  transfer. *See* knowledge sharing
  translators 107
  use 47–48
  workers 107
Knowledge Flow Analysis model
    58–62
knowledge-intensive organisation
    40
Knowledge Life Cycle 65
knowledge management
  activities 41–53
  challenges 15–38
  cycle 63–66
  definitions 9–10
  during the event 176–178
  frameworks 53–66
  objectivist perspective 66, 70
  post-event 178–180

practice-based perspective 71–93
pre-event 173–176
relational 72–79, 143, 177
roles and responsibilities 106–108
three generations 11
knowledge management system,
    basic model 53
Knowledge Mapping 44
knowledge practices 73
  collaboration 127
  role of emotions 78–79
  formal and informal rituals 74–75
  pre, during and post event 172–173
knowledge sharing/ transfer 48–51
  Appreciative Sharing of
    Knowledge 160–163
  cultural factors 119
  defined 48–49
  during the event 177
  through storytelling 163–167
  ICT support 50
  inter-disciplinary teams 51
  motivation 124–125
  rituals 75
  trust 125–126
  volunteers 26–27, 51
Knowledge Sharing Systems 44

leadership
  in learning organisations 108
  leading-by-example 108, 177
learning
  by being told 45–46
  by doing 73, 177
  by observation 45–46
  communities-of-practice 83
  from mistakes 78
  from projects 20
  on-the-job 98
  organisational 7–8, 167, 173
  preferences and personal
    characteristics 47
  re-learning 33–34
  unlearning 29

McElroy
  Distributed Organisational
    Knowledge Base 64
  knowledge management cycle
    63–66
  event example 64–65
memory. *See* organisational memory
mentoring 73, 175
middle managers
  as knowledge creators 101
middle-up-down management
    101–102, 174
motivation
  extrinsic 123–124
  for sharing knowledge 51, 123–125
  human resource management 96
  intrinsic 123–124
  lack of 24
  recognition and reward 99

Nonaka and Takeuchi
  four modes of knowledge
    conversion 54–57
  critique 57–58
  event example 57
  spiral evolution of knowledge
    creation 56
  top-down, bottom-up and middle-
    up-down management 100–102

objectivist perspective on
    knowledge management 70
open office design
  surveillance or sharing 137
operational knowledge 28–29
organisational capital 40
organisational culture 113–120, 173
  categories 117
  factors that affect knowledge
    processes 119

in academic literature 119
Queensland Music Festival 120–122
organisational identity 114–116, 173
  characteristics 115–116
organisational knowledge 7–8, 141
organisational learning 7–8, 167, 173
organisational memory 7–8, 21, 179
  knowledge storage 51–52
organisational structure 95, 174
  decentralised 103
  flat 101, 145
  hierarchical 80, 100–102, 135, 138–139
  middle-up-down management 101
organisational vision 116, 173
  Queensland Music Festival 120–122

participation, in communities-of-practice 81–83
performance assessing, post-event 178–179
politics in organisations 136–137
post-event evaluation 16
power 134–154
  as a resource 137–142
  at organisational level 141
  bases of 140
  expert knowledge 137
  hierarchical 138–139
  in academic literature 135
  in communities-of-practice 86
  legitimate 138–140
  misuse 139, 141–142
  politics and conflict in organisations 136–137
  positive 144–147, 178
  rules of the game 117, 134–135
power/knowledge
  concept 142–144
  festival example 146

Foucault 143
practice-based perspective on knowledge management 71–93
project-based organisations
  film industry 20–21
  knowledge management challenges 19–21
  project-learning 20
  reinventing the wheel 33
pulsating organisations 17–19, 60, 172
  knowledge management challenges 18–19, 96–99
  repeat pulse event organisation 18, 33
  single pulse event organisation 18

Queensland Music Festival
  interdisciplinary pods 104–105
  knowledge sharing ritual 76–78
  power and empowerment 147–151
  vision, identity and culture 120–122

recommendations 171–180
  during the event knowledge management 176–178
  post-event knowledge management 178–180
  pre-event knowledge management 173–176
recruitment 26, 97
reification 81
reinventing the wheel 16, 33–34
relational knowledge management 72–79, 143, 177
re-learning 33–34
research opportunities 180–181
resources, for knowledge management activities 175
  lack of 22–23
rituals 74–75, 117, 176–177

formal 74
informal 75
Queensland Music Festival 76–78
rules of the game 117, 134–135

shared understanding 114–116
  group knowledge 7
skills 21–22, 25–27
  development 98–99
  lack of 24, 96
social capital 40
soft factors in acquiring knowledge 118
staff development 97–99
  job rotation 99
  on-the-job learning 98
staff productivity 97
staff retention 97, 99–100, 179–180
  knowledge leakage 99
stories 164–165
  of success 143, 162, 165–166, 179
storytelling
  for knowledge sharing 163–166
  importance in knowledge management, 167
strategic knowledge 28
structure. *See* organisational structure
surveillance 136

tacit knowledge 5–6
  cognitive dimension 5
  technical dimension 5
time, lack of 22–23
time outs during events, to recharge 178

timing in projects 23
top-down management 100
training 46, 98
  allocating resources 175
  on-the-job 24–26
  time constraints 24
trust
  affect-based 126–127
  cognition-based 126–127
  defined 125
  enabler for knowledge sharing 125–127
  lack of 24, 125

unlearning 29

vision. *See* organisational vision
volunteers
  knowledge sharing 26–27, 51
  lack of skills and knowledge 26
  large number of 25–27
  motivation 51, 123
  recruitment and selection 26
  retention 27, 51, 124–125, 179–180
  sense of community 176
  training 175
  trust 25

Wiig
  four knowledge flow dimensions 61
  Knowledge Flow Analysis 58–62
  event example 61–63
work groups
  and communities-of-practice 80

Printed by Printforce, United Kingdom